IN THE PURSUIT OF
REAL SCIENCE
IN A
MODERN-DAY SOCIETY

DALE J. BLOCK

ISBN: 979-8-89633-051-6 (sc)
ISBN: 979-8-89633-052-3 (e)

PAGE
SOLUTIONS
Page Solutions
541 Buttermilk Pike
Crescent Springs, KY 41017

Printed in the United States of America

CONTENTS

PREFACE

Patton (2018) defined *Science* based on the Oxford English Dictionary in the following manner:

1. ... intellectual and practical activity encompassing the systematic study of the structure and behavior of both the physical and natural world through observation and experimentation...
2. ... systematically organized body of knowledge on a particular subject...

Merriam-Webster's Dictionary (2024) defines *Science*:

1. ... systemized knowledge derived from observation, study, and experimentation carried on in order to determine the nature of principles of what is being studied...

These definitions of science as a systemized structure and orderly processes that delivers an outcome in the world around us does not speak to the true essence of what is real science. Carl Sagan, the renowned astronomer, astrophysicist, and science communicator, in his book, *The Demon-Haunted World: Science as a Candle in the Dark* (1995) defined science this way,

... is a way of thinking much more than it is a body of knowledge.

Sagan believed that science is not just about accumulating facts but about cultivating a mindset that values critical thinking, logical reasoning, skepticism, and a systematic approach to understanding the world and beyond, i.e., real science. Sagan emphasizes the importance of scientific thinking in everyday life and its role in combating superstition and pseudoscience.

Real science is a human invention, a social construct, that is not attributed to a specific author but is rather a paraphrase or summary of ideas that have been discussed by various scholars, philosophers, and sociologists of science. These ideas are rooted in the field of real science and technology studies (STS) and the sociology of knowledge. Thomas Kuhn in his influential book *The Structure of Scientific Revolutions* (1962), argued that scientific knowledge is shaped by paradigms, which are frameworks of understanding that guide research. These paradigms are subject to change based on social, historical, and cultural factors, leading to scientific revolutions. Bruno Latour and Steve Woolgar in their book *Laboratory Life: The Construction of Scientific Facts* (1979), explored the social processes involved in the production of scientific knowledge, emphasizing that what we consider scientific facts are the result of social and cultural practices within scientific communities. Karl Popper, though he did not explicitly label real science as a social construct, highlighted the human elements inherent in the scientific process through his philosophy of science, particularly in his 1959 work on falsifiability. Finally, David Bloor, a key figure in the sociology of scientific knowledge (SSK), argued in his book, *Knowledge and Social Imagery* (1976), that scientific knowledge is influenced by social factors and should be studied as a socially constructed phenomenon.

The nature of scientific knowledge (Whewell, 2023), emphasizes that real science is not just a collection of objective facts but is shaped by the social and human contexts in which it is developed. It also highlights the idea that real science, as we understand and practice it, is a product of human thought, culture, and social organization.

Real science as a human invention is therefore created by humans (Restivo et al., 2000). Real science is not something that exists independently in nature. It is a method and framework created by humans to understand, explain, and manipulate the natural world. The tools and techniques of real science, such as experiments, hypotheses, and peer review, are human-made processes designed to systematically investigate the world. Real science evolves over time as human understanding grows and changes. New discoveries, technologies, and paradigm shifts continuously reshape our understanding of the natural and physical world.

Real science as a social construct is heavily influenced by society (Cole, 1992). The development of real science is influenced by the culture, values, and norms of the society in which it is practiced. Different societies may prioritize different areas of scientific inquiry based on their needs, interests, and resources. Scientific knowledge is produced through a collaborative process involving communities of scientists who work together, share ideas, and build upon each other's work. This makes real science a social activity, dependent on communication, collaboration, cooperation, coordination, and consensus. The direction, funding, and application of scientific research are often influenced by outside stakeholders driving social, political, and economic forces. For example, research priorities may be shaped by government policies, public interest, or corporate investment.

Recognizing real science as a human invention and social construct encourages us to understand it in its historical, cultural, and social context. It helps us see how scientific knowledge is shaped by the people and societies that produce it. This perspective invites critical reflection on the assumptions, biases, and power dynamics that can influence scientific practice. It reminds us that real science is not entirely objective or neutral but is shaped by human interests and values. Understanding real science in this way emphasizes the responsibility of both scientists and society to ensure that scientific knowledge is used ethically and for the common good. While real science is a powerful tool for understanding the world around us, it is also a product of human invention and social interaction, with all the complexities and implications that come with it.

The distinction between *Real Science* and *Science* often hinges on the context in which these terms are used, particularly when discussing the integrity and rigor of scientific inquiry. The term, *Science*, generally refers to the systematic pursuit of knowledge through observation, experimentation, and the application of the scientific method. It encompasses a broad range of disciplines, including physics, chemistry, biology, and social sciences. Science relies on empirical evidence, logical reasoning, and peer review to build reliable knowledge about the natural world. *Real Science* is used to emphasize science that adheres strictly to the principles of objectivity, empirical evidence, and rigorous methodology. It can imply a critique of practices or findings that may be labeled as *science* but lack these qualities. *Real Science* suggests a commitment to integrity, transparency, and a rejection of pseudoscience, bias, or research influenced by external pressures such as funding sources, political agendas, or media influence.

While *Science* is a broad term for the study and understanding of the natural world, *Real Science* underscores a stricter adherence to the principles of honest, unbiased, and rigorous scientific practice.

Existential threats to human survival (e.g., climate change, pandemics, natural disasters, etc.) are getting worse, not better (Sears, 2020). Much of this is the direct result of human ignorance and annoyance. This recent phenomenon is best stated as, *A stick your head in the sand and hope someone else can make our natural world better*. That is not working either.

Do we really need another book on the basics of pursuing real science in a modern-day society? The answer is a resounding yes! Several driving forces are contributing to the gradual decline of human society as we know it. In a world where misinformation and pseudoscience spread rapidly, it's essential to provide clear, accurate explanations of what real science is and how it works. Doing so helps dispel myths and fosters a better understanding of the scientific method. Additionally, a book that promotes critical thinking, logical reasoning, and skepticism can empower readers to critically evaluate evidence and differentiate between credible sources and unreliable ones. This is especially vital in an era where people are constantly bombarded around-the-clock with polarizing and conflicting information.

Real science plays a fundamental role in addressing global challenges like climate change, emerging infectious diseases, noncommunicable chronic conditions, and rapid cycle technological advancement. A book exploring how real science contributes to solving these issues and why it's important to support and trust the scientific community is needed today with hopes for creating a sustainable tomorrow for the next generations of scientists.

Today, more than ever before in history, skepticism and distrust are on the rise, often driven by political agendas or misinformation. A book that tackles this issue by explaining how scientific consensus is achieved within the scientific community and why it deserves trust could help the public better understand its importance. Furthermore, by highlighting the pursuit of real science, such a book could inspire young people to pursue careers in science, technology, engineering, and mathematics (STEM), thereby contributing to future innovations and advancements.

This book also documents the challenges that modern scientists encounter today, including funding cuts, political interference, and public misunderstanding, while showcasing the resilience and dedication of those devoted to the scientific pursuit. Real science is not only about discovery but also about ethical responsibility. This book delves into the ethical considerations in scientific research, emphasizes the importance of integrity, and examines the consequences of unethical practices.

The primary goal of writing this book is to cultivate a more informed and scientifically literate society, one that can make decisions benefiting both individuals and the greater good of the public.

Dale J. Block, MD, MBA

March 2025

CHAPTER 1

Introduction:
The Essence of Real Science.

The essence of real science lies in its dedication to understanding the natural world through systematic inquiry, empirical evidence, and critical thinking (Quoc et al., 2022). Driven by curiosity and the pursuit of real knowledge, real science relies on observation, experimentation, and rigorous hypothesis testing. It upholds values such as transparency, reproducibility, and peer review, ensuring that findings are credible and independently verifiable. Real science is marked by a willingness to revise its conclusions; it adapts and evolves as new evidence emerges, continuously refining its understanding of the world around us. At its core, real science seeks to uncover truths, even when they challenge existing beliefs or assumptions, with the ultimate aim of advancing human knowledge and improving the quality of life for all, and not just a select few.

1.1 Definition and Significance of "Real Science".

1.1.1 True Definition of Real Science.

Real science is the disciplined pursuit of knowledge, rooted in the systematic study of the natural world through observation, experimentation, and critical analysis (Patton, 2018; Oxford Reference, 2024). It is defined by a methodical approach that begins with curiosity-driven questions, followed by the formulation of hypotheses that can be tested and potentially falsified. The process involves collecting empirical data, analyzing results, and drawing conclusions based on evidence generated through experimentation and observation. Real science is not static; it is dynamic and self-correcting, with findings subject to scrutiny, peer review, and replication. This iterative process ensures that scientific knowledge is valid and reliable and constantly refining our understanding of the universe. At its core, real science is guided by the principles of skepticism, critical thinking, logical reasoning, transparency, and truthfulness, seeking to uncover facts that are universally applicable and independent of personal beliefs or biases.

1.1.2 Why is It Important to Pursue Real Science in a Modern-Day Society?

Pursuing real science in modern society is critical (Nielsen, 2020; Collins et al., 2017) for several reasons:

1. *Advancement of Knowledge*: Real science helps us understand the natural world, leading to new discoveries and innovations that can improve our quality and quantity of life.
2. *Technological Progress*: Scientific research drives technological advancements, which can enhance individual and population-based productivity, collaboration, health, wellness, and well-being along with other aspects of daily life.
3. *Problem-Solving*: Real science provides tools and methodologies to build capacity to address and solve complex problems, such as climate change, pandemics, and supply chain inefficiencies.
4. *Informed Decision-Making*: Scientific knowledge allows policymakers and the public-at- large to make informed decisions based on evidence rather than speculation or misinformation.
5. *Economic Growth*: Scientific research can lead to the development of new industries and job opportunities, contrib-

uting to economic growth, workforce development, stability, and sustainability.

6. *Health and Medicine*: Advancements in medical and health system science have significantly enhanced health outcomes, leading to increased life expectancy and bringing health span closer to life span.

7. *Education and Critical Thinking*: Studying real science fosters critical thinking, logical reasoning, embraces uncertainty, and refines the ability to analyze and interpret data—essential skills in today's information-rich, always-connected society.

8. *Global Challenges*: Many of the challenges we face today, such as economic inequity, climate change, institutional bias, and emerging infectious diseases require a rigorous approach world-wide with disciplined scientific inquiry for effective solutions.

In essence, real science is foundational to the continued progress, capacity-building, and sustainability of a modern-day society.

1.1.3 The Primary Goal of Real Science.

The primary goal of real science is to expand knowledge, which consists of the facts, information, and skills an individual gains through experience and education (Britt et al., 2014). In contrast, the primary goal of technology is to apply this acquired knowledge.

Real science focuses on abstract knowledge—generalizations, rules, or concepts that are not directly related to real-life actions or objects. These thoughts and ideas are created by combining concepts and experiences to form knowledge, resulting from developing theories based on patterns, relationships, and general information. The design and development of technology then take this abstract knowledge and transform it into material structures and processes, propelling modern society forward.

The fundamental components of real science and technology include knowledge, skill building, discovery, understanding, and application. The first four components are established with scientific inquiry. Application is then taking the results of scientific inquiry and applying those results to design technology. The foundation for real science comes from this back-and-forth between real science and technology.

Achieving the fundamental objectives of real science and technology necessitates specific frameworks and methodologies. In real science, this entails formulating insightful questions to obtain valid, reliable, and reproducible answers, a process commonly referred to as discovery science.

Discovery science enhances our understanding of the natural world and fosters creativity and innovation across scientific disciplines. It leads to technological advancements, biological and medical breakthroughs, and environmental adaptation and mitigation solutions to sustain high-level ecosystem services. By organizing structures and processes, discovery science enables informed decision-making in science education, research, and practice, which in turn drives the development of effective public policies (Davidson, 2015; Fortunato et al., 2018).

Real science does have its limitations. First, it only addresses what can be observed. Second, observations can be flawed. Third, institutional bias remains a persistent and unresolved challenge. Fourth, real science cannot make value judgments. Fifth, it cannot offer universal statements. Finally, real science cannot provide definitive answers.

The above limitations of real science set the stage for the scientific community to be thoughtful and deliberate when going outside of their comfort zone and enter into the space of the public-at-large to communicate important findings from observation and experimentation.

1.1.4 Significance of Real Science.

The significance of real science extends far beyond the confines of laboratories and academic institutions; it is fundamental to the progress and well-being of society (Alexandrova, 2017; Ryff, 2018). By providing a reliable framework for understanding the world, real science enables us to make informed decisions, solve complex problems, and innovate creatively across all areas of human endeavor. In medicine, real science drives the discovery of new treatments and technologies that save lives and improve health outcomes (NRC, 2011). In environmental stewardship (Bennett et al., 2018), real science helps us understand and mitigate the adverse impact of human activity and greenhouse gas (GHG) emissions on the planet. The scientific method fosters critical thinking, logical reasoning, and problem solving all the while cultivating a mindset that values fact over opinion, helping individuals and societies navigate the complexities of the modern natural world. Moreover, real science promotes collaboration and shared knowledge, uniting scientists and the public-at large around-the- world and across cultures and disciplines in the common quest for truth. Its impact is profound, laying the foundation for technological advancements, improving quality of life, and empowering humanity to address the challenges of today in order to have a better tomorrow.

1.2 The Role of Skepticism and Critical Thinking in Real Science.

1.2.1 Role of Skepticism in Real Science.

Skepticism plays a vital role in real science, serving as a safeguard against error, bias (i.e., both implicit and explicit), and misinformation (Rutjens et al., 2021). It involves questioning long-standing conventions, critically and logically evaluating empirically-derived evidence, and maintaining a vigilant approach to accepting new scientific findings.

In real science, skepticism isn't about cynicism or doubt for its own sake; it's a valuable tool that drives the quest for truth. By critically examining claims and rigorously testing hypotheses, skepticism ensures that scientific conclusions rest on solid, reproducible evidence rather than on unfounded opinions or unverified data. This process of questioning and verification is essential for advancing knowledge and safeguarding against the acceptance of false or misleading information. Skepticism also encourages scientists to remain open-minded, acknowledging that current understanding can always be revised in light of new evidence. In this way, skepticism propels scientific progress by fostering intellectual discourse and preventing complacency.

1.2.2 Role of Critical Thinking in Real Science.

Critical thinking is the cornerstone of real science, enabling scientists to analyze information, draw logical conclusions, and solve complex problems (Santos, 2017). It involves the ability to assess the validity of arguments, identify logical fallacies, and synthesize diverse pieces of information into clear, evidence-based conclusions.

In real science, critical thinking allows researchers to navigate vast amounts of data, identify meaningful patterns, and distinguish significant and reliable findings. It also guides scientists in designing rigorous experiments, accurately interpreting results, and understanding the broader implications of their work. Critical thinking nurtures creativity and innovation by encouraging scientists to look beyond the obvious, question established paradigms, and explore new possibilities. By applying critical thinking, scientists ensure that their work is both methodologically sound and impactful. Ultimately, critical thinking empowers real science to advance knowledge and solve real-world problems, making it an essential tool in the scientific process.

1.3 Overview of Goals and Objectives for Exploring the Concept of Real Science.

In the Pursuit of Real Science in a Modern-Day Society aims to provide the reader with a comprehensive exploration of the concept of real science, delving into its fundamental principles, methodologies, and the critical role it plays in society. The objectives are to define and clarify what constitutes real science, differentiate it from pseudoscience, and emphasize the importance of skepticism, critical thinking, logical reasoning, and ethical practices in scientific inquiry and problem solving. This book also seeks to equip its readers with a deeper understanding of how real science works, why it is essential for global human progress, and how to apply scientific thinking in everyday life.

The beginning of the book introduces the definition and concept of real science, outlining its significance and why it remains a cornerstone of modern civilization. It sets the stage for the detailed exploration of scientific principles and methodologies that follows.

In Part 1, The Foundations of Real Science, the reader is introduced to the scientific method, the role of evidence in real science, and the evolution of scientific theories. This part also delves into the history and philosophy of real science, discussing the origins of the scientific method and the evolution of scientific thought. It also addresses the criteria that distinguish real science from other forms of inquiry. It provides a detailed examination of the scientific method, including the formulation of hypotheses, experimentation, data collection, and analysis. It explores the importance of empirical evidence, repeatability, and peer review in ensuring the reliability of scientific findings.

In Part 2, the Challenges of Modern Science, the discourse begins by examining the characteristics of pseudoscience and its impact on shaping public opinion. It delves into the differences between real science and pseudoscience, providing examples of common misconceptions. The discussion also addresses the longstanding influence of bias, both implicit and explicit, within the scientific community, the challenges of inadequate funding for researchers, and the current reproducibility crisis in real science.

In Part 3, the Ethics of Real Science, chapters on scientists' responsibility to the public, the intersection of real science and society, and the role of real science in education will be presented. The ethical dimensions of scientific research are explored here, covering topics such as research integrity, the responsible use of technology, and the societal implications of scientific discoveries. The discourse emphasizes the importance of conducting real science in a way that benefits humanity while minimizing harm. The role of real science in education examines best practices for educating students in real science and expectations of what students should know about science prior to starting post-secondary education.

In Part 4, the Future of Real Science, will cover the role of technology in advancing real science, the importance of global scientific collaboration, and the measures taken to ensure scientific integrity. It will explore the broader implications of scientific advancements, including their impact on technology, medicine, and the environment, as well as the crucial role of real science in addressing global healthcare delivery system challenges and enhancing both the quantity and quality of life.

The Conclusion reinforces a call to action, urging affected stakeholders world-wide to commit to the pursuit of real science. It encourages readers to integrate the principles of scientific thinking into their daily lives, emphasizing the enduring importance of real science and the need for steadfast adherence to scientific principles in our increasingly complex and interconnected world. The final chapter of the book emphasizes the value of skepticism, critical thinking, logical reasoning, and ethical responsibility in the continuous quest for knowledge.

PART I

The Foundations of Scientific Inquiry.

The foundations of scientific inquiry are built on principles of observation, experimentation, and empirically-driven evidence-based reasoning. Scientific inquiry begins with careful observation and the formulation of questions about natural phenomena. These questions lead to hypotheses, which are testable predictions about how variables interact. Scientists test these hypotheses by conducting highly-structured, controlled experiments and systematically collecting data to gather evidence. The analysis of this evidence allows for the development, refinement, or rejection of theories. Central to scientific inquiry is the emphasis on validity, reliability, reproducibility, peer review, and openness to new evidence, ensuring that scientific knowledge evolves based on rigorous investigation and critical evaluation.

CHAPTER 2

The Scientific Method - A Time-Tested Approach.

The scientific method is a time-tested approach that forms the cornerstone of scientific inquiry (Nuzzo, 2014; Poincare, 2022). It is an organized, highly-structured, and logical process used to explore observations, answer questions, and solve problems. The method typically involves making observations, forming a hypothesis, conducting experiments, collecting and analyzing data, and drawing conclusions. This iterative process allows scientists to build on previous knowledge, test ideas under highly-structured and controlled conditions, and refine or reject hypotheses based on empirical evidence. The scientific method, with its emphasis on reproducibility and peer review, ensures that findings are reliable and enhance our broader understanding of the natural world. Its rigorous structure has made it an essential tool for advancing knowledge across all scientific disciplines.

2.1 What is Real Science?

As discussed in the preface and previous chapter, *real science* is a systematic enterprise that builds and organizes knowledge in the form of testable explanations and predictions about the universe. It is both a body of knowledge and a process of inquiry. The primary goal of real science is to understand and explain natural phenomena in a systematic and objective way (The Science Council, 2024).

Scientific knowledge is based on empirical evidence, which is gathered through observation and experimentation. It relies on data that can be measured, observed, and tested. Scientists use systematic methods of observation and experimentation to gather data. This involves following a highly-structured, organized approach to ensure validity, reliability, and reproducibility in the results.

Scientific theories, models, and frameworks are often designed and developed to make predictions about future observations or experiments. The ability to make accurate predictions about future investigations is a hallmark of successful scientific theories.

Scientific findings should be replicable by independent researchers. Reproducibility is essential for validating scientific claims and ensuring the reliability of results. Scientific theories must be testable and open to being proven false based on empirical evidence. This distinguishes scientific ideas from unfalsifiable or untestable claims (e.g., pseudoscience claims).

Real science is a cumulative and a self-correcting process. New findings build upon existing knowledge and incorrect or incomplete ideas are revised or discarded as new empirical evidence emerges.

Scientific inquiry is carried out without preconceived notions. Impartiality is crucial for ensuring that the results of scientific investigations are as valid and reliable as possible. Scientific research undergoes peer review, where subject matter experts in the same field and sometimes, in other fields, critically evaluate the methods, results and conclusions of a study before it is published. This process helps maintain the quality and integrity of scientific knowledge.

Real science encompasses a variety of disciplines. Each scientific discipline follows the scientific method to systematically investigate and comprehend specific aspects of the natural world. The scientific method involves making observations, formulating hypotheses, conducting controlled experiments, collecting data and drawing conclusions based on the empirical evidence generated by following the scientific method.

2.2 How Does Real Science Work?

Scientific inquiry operates through a systematic and methodological approach known as the scientific method (Eastwell, 2010). Although the specific steps and details may differ between academic disciplines, the general process includes the following key stages:

1. *Observation*: Scientists begin by making observations about the natural world. These observations can be qualitative (descriptive) or quantitative (measurable). Observations hopefully lead to questions or just basic curiosity about a specific phenomenon.

2. *Question*: Based on observations, scientists formulate specific questions or hypotheses. A hypothesis is a testable statement that predicts the relationship between variables.

3. *Research*: Before conducting experiments or further investigations, scientists review existing literature (e.g., peer-review journals, gray literature) and ongoing research to understand what is already known about the topic to help inform the design and development of their study.

4. *Hypothesis*: Scientists propose a hypothesis that can be tested through experimentation or observation. The hypothesis often includes an explanation of the expected outcome.

5. *Experimentation or Observation*: Scientists design, develop, and implement experiments or make systematic observations to gather data. Experiments are highly-structured investigations with controlled variables, while observations involve collecting data from natural settings. Data can be qualitative or quantitative and is crucial for drawing conclusions.

6. *Analysis*: Scientists analyze the collected data using statistical or other methods to identify patterns, trends, or

relationships. The analysis helps determine whether the results support or disprove the hypothesis.

7. *Conclusion*: Based on the analysis, scientists draw conclusions regarding the original hypothesis. They assess whether the empirical evidence generated supports the proposed explanation or if further investigation is needed.

8. *Communication*: Scientists communicate their findings to the scientific community through research publications and scientific presentations. This allows the scientific community (i.e., peers) to review, replicate, and build upon the research. Peer review is a critical part of the scientific process, where experts in the field assess the validity, reliability, and reproducibility of the study.

9. *Revision and Replication*: Real science is self-correcting. If new evidence contradicts or modifies existing theories, scientists revise their understanding of the topic. Replication of experiments by other researchers is essential to confirm the validity of findings and ensure the reliability of scientific knowledge.

10. *Theory Building*: Over time, successful hypotheses and findings contribute to the development of scientific theories. A scientific theory is a well-substantiated explanation of some aspect of the natural world that is supported by a large body of empirically-derived evidence.

This highly comprehensive and iterative process of observation, questioning, experimentation, and results analysis is the foundation of scientific methodology. It allows scientists to refine their understanding of the natural world and contributes to the continuous advancement of scientific knowledge.

2.3 What Makes Real Science Unique from Other Ways of Investigating the World Around Us?

Real science is unique from other ways of examining the world around us due to its distinguishing processes, methodologies, frameworks, theories, laws and principles (IOM, 1992, IOM, 1993).

Real science relies on observed and experimental evidence and emphasizes the importance of objective, measurable data as the basis for understanding the natural world. The scientific method provides an organized and comprehensive approach to inquiry. It involves making observations, forming hypotheses, conducting experiments, and analyzing results. This highly-structured process helps ensure precision, clarity, and impartiality in scientific investigations. Scientific hypotheses, theories, and principles must be transparent and testable to being confirmed incorrect through empirical evidence. This differentiates scientific ideas from beliefs, claims, and opinions that cannot be objectively tested or proved.

Scientific findings should be replicable by independent researchers. Reproducibility is a key aspect of scientific inquiry, as it allows others to authenticate and corroborate results, contributing to the trustworthiness of scientific knowledge. Scientific theories often have predictive power, meaning they can make truthful predictions about upcoming observations or experiments. This ability to anticipate outcomes adds strength to scientific explanations.

Real science aims for impartiality in the design, development, implementation, and interpretation of experiments. The scientific community values impartiality and strives to reduce personal or cultural bias on results. Scientific research undergoes peer review by the scientific community, where subject matter experts in the same field critically analyze the methods, results, and conclusions before journal publication. Fidelity to this process helps maintain the quality and truthfulness of scientific information and knowledge.

Scientific knowledge is cumulative and builds upon previous discoveries. New findings contribute to existing theories and principles, expanding our comprehension of the wonders of the natural world. This characteristic distinguishes real science from more static or inflexible approaches. Scientific principles and laws are often universal. They apply consistently across different situations and conditions and are not restricted by cultural, geographic, or individual variations.

Real science is guided by ethical principles, emphasizing integrity, impartiality, responsibility, accountability, truthfulness, and transparency. Ethical considerations are foundational in conducting research and reporting results dutifully. Scientists often collaborate, cooperate, coordinate, and communicate with peers, sharing information and expertise. Working with one's peers enhances the diversity of viewpoints and proficiency, contributing to stronger and more wide-ranging scientific investigations of the natural world. These distinct characteristics set real science apart as a rigorous, empirically-driven, evidence-based, and self-correcting approach to understanding the natural world. While other methods of exploring the world—such as philosophy, religion, or personal experience—provide valuable insights, they often lack the practical and systematic foundations that define scientific inquiry.

2.4 Where Does Real Science Begin and Where Does It End?

The boundaries of real science are not always sharply defined, and the demarcation between what is considered real science and what is not can be a subject of philosophical and social debate (IOM, 1992, IOM, 1993). However, there are certain characteristics that help delineate the scope and scale of scientific inquiry.

Real science typically involves empirical investigation, relying on observation, measurement, and experimentation to gather evidence. If a field or approach lacks a basis in empirical evidence, it may fall outside the realm of traditional science. Scientific ideas are expected to be testable and open to falsification. If a hypothesis, theory, or principle cannot be subjected to empirical testing, it may be considered less scientific. This criterion distinguishes real science from unfalsifiable claims.

Real science follows an orderly progression, often referred to as the scientific method or inquiry. This involves formulating hypotheses, conducting experiments, and analyzing data. Disciplines or approaches that lack this organized and methodical approach may not be classified as real science.

Real science emphasizes objectivity in the pursuit of information and knowledge, seeking to minimize personal partiality and bias. If an approach is heavily influenced by personal beliefs, ideologies, or lacks objectivity, it may be considered less scientific.

Scientific findings are expected to be reproducible by independent researchers. If a particular experiment or phenomenon cannot be reliably reproduced, it may raise questions about the scientific legitimacy of the results.

The boundaries of real science can be dynamic and may evolve over time. Interdisciplinary fields, such as astrobiology or bioinformatics, highlight the fluid nature of scientific inquiry. In addition, scientific disciplines may overlap with other approaches of inquiry, such as philosophy, especially in areas where empirical evidence is limited or difficult to attain. Discussions about the limits of real science often intersect with philosophical questions about the nature of knowledge, distinguishing real science from non-science, and the relationship between real science and other fields of critical study, such as art, ethics, or religion.

While real science provides a powerful and reliable method for understanding the natural world, it may not address all questions or aspects of the human experience. Other forms of inquiry, such as philosophy, ethics, and the arts, contribute valuable perspectives and insights that complement the scientific understanding of the natural world. The boundaries of real science are therefore best understood as dynamic and context-dependent rather than fixed or absolute.

2.5 Which Kinds of Activities Count as Being Scientific?

Activities are considered scientific if they adhere to the principles and methods of scientific inquiry. As previously stated, scientific activities involve the highly-structured and organized observation of natural phenomena. Scientific activities follow a systematic approach, i.e., the scientific method. This includes making observations, forming hypotheses, conducting experiments, and analyzing data (IOM, 1992, IOM, 1993).

Scientific activities involve hypotheses or theories that are testable and open to falsification. They produce results that can be replicated by independent researchers and aim for objectivity, minimizing personal and institutional bias in the design, conduct, and interpretation of experiments. These activities often lead to hypotheses or theories with predictive power and are subject to peer review, where experts evaluate the methods, results, and conclusions before publication. Scientific activities contribute to the gradual accumulation of knowledge and are conducted with a commitment to ethical principles, including honesty, integrity, and transparency in research.

Some examples of scientific activities include conducting controlled experiments in a laboratory to test a hypothesis, observing and collecting data on natural phenomena in the field, analyzing data using statistical methods to identify patterns or relationships, developing mathematical models to explain observed phenomena, and formulating and testing hypotheses to understand cause-and-effect relationships.

While these characteristics are common in many scientific activities, the application of the scientific method can vary across disciplines. Different fields, such as physics, biology, psychology, and sociology, may have distinct methods and approaches that align with their subject matter. Interdisciplinary fields and emerging areas of study may also exhibit unique characteristics within the broader framework of scientific inquiry (Godfrey-Smith, 2003).

2.6 Moore's Criteria

The iconic professor of biology, John Moore, spent much of his career writing and speaking on the history of the natural sciences. His 1993 book, *Science as a way of knowing: The foundations of modern biology*, was no exception. Moore created a list of criteria for determining whether a certain activity qualifies as "science" within this context: "Is biology an autonomous science?" It can be rephrased into two parts: "Is biology, like physics and chemistry, a science?" and "Is biology a science exactly like physics and chemistry?" The following is a list of criteria created and published by Dr. Moore (1993):

1. A science must be based on data collected in the field or laboratory by observation or experiment, without invoking supernatural factors.
2. Data must be collected to answer questions, and observations must be made to strengthen or refute conjectures.
3. Objective methods must be employed in order to minimize any possible bias.
4. Hypotheses must be consistent with the observations and compatible with the general conceptual framework.
5. All hypotheses must be tested, and, if possible, competing hypotheses must be developed, and their degree of validity (problem-solving capacity) must be compared.
6. Generalizations must be universally valid within the domain of the particular science. Unique events must be explicable without invoking supernatural factors.
7. In order to eliminate the possibility of error, a fact or discovery must be fully accepted only if (repeatedly) confirmed by other investigators.
8. Science is characterized by the steady improvement of scientific theories, by the replacement of faulty or incomplete theories, and by the solution of previously puzzling problems.

Mayr (1997), within the context of Moore's eight criteria, provided his own criteria for determining whether a certain activity qualifies as "science":

> *... yes, biology is like physics and chemistry, a science. But biology is not a science like physics and chemistry; it is rather an autonomous science on par with the equally autonomous physical sciences ... tasks of the philosophy of biology is to establish what the common features are which biology shares with the other sciences, not only in methodology but also in principles and concepts.*

2.7 Real Science is Not Religion.

Real science and religion are fundamentally different in their methodologies, goals, and underlying principles, despite both seeking to understand aspects of human existence and the universe (Ferngren, 2022).

Real science is grounded in empirical evidence and the scientific method. The comprehensive and iterative process of testing and refining knowledge allows real science to progressively build a more accurate understanding of the natural world.

In contrast, religion is based on faith, spiritual experiences, and sacred texts. It provides moral guidance, meaning, and a sense of purpose, often involving beliefs that are not subject to empirical testing or falsification. Religious doctrines are typically considered undisputable truths by devotees, derived from divine revelation or long-standing tradition. While religious beliefs can evolve, the process is often influenced by theological interpretation rather than empirical evidence. Religion addresses questions of purpose, ethics, and the nature of existence in ways that are deeply personal and culturally specific, offering a framework for understanding aspects of life that science does not address.

One key distinction between real science and religion is that real science does not prescribe moral or ethical values; it seeks to describe and explain phenomena without making value judgments. The results of scientific inquiry can inform ethical discussions and policy decisions, but real science itself remains neutral. Religion, on the other hand, often provides comprehensive moral frameworks and ethical guidelines for devotees. These frameworks can offer solace, community, and a sense of identity, all playing a central role in shaping cultural and individual values.

Furthermore, the nature of evidence and proof differs significantly between real science and religion. Scientific claims require empirical support and must be reproducible by independent researchers. This reliance on observable and measurable phenomena means that scientific knowledge is always provisional and open to challenge. Religious beliefs, however, are typically based on spiritual conviction and faith, which do not require empirical evidence and are often considered beyond the realm of scientific scrutiny. This difference in the basis for belief highlights the distinct epistemological foundations of real science and religion.

In summary, real science and religion operate in fundamentally different domains: real science in the realm of empirical investigation and explanation, and religion in the realm of faith, meaning, and moral guidance. While both are valuable and offer unique insights, they are not interchangeable and serve different purposes in human understanding and experience. Recognizing these differences allows for a more nuanced appreciation of how each contributes to the broader tapestry of human knowledge, culture, and values.

2.8 Real Science vs History: Distinct Yet Complementary Fields of Study

Real science and history are distinct fields of study with different methodologies, objectives, and focuses, yet both contribute significantly to our understanding of the natural world (Matthews, 2024). As previously presented, real science is primarily concerned with uncovering the principles and laws governing the natural world through observation, experimentation, empirical evidence, and the scientific method. In contrast, history is dedicated to understanding the human past, focusing on events, cultures, and social dynamics through the analysis of historical records, artifacts, and narratives.

As previously discussed, the methodology of real science revolves around the scientific method. This process emphasizes objectivity, predictability, and repeatability. Scientific knowledge is cumulative and self-correcting, with new discoveries often building upon or revising previous understandings. For example, the development of the theory of evolution (Sober, 2024) or the laws of physics (Chen, 2025) illustrates how scientific inquiry advances our knowledge of natural phenomena through rigorous testing and validation.

In contrast, history relies on qualitative analysis and interpretative methods to reconstruct and understand past events (Rodney, 2024). Historians use primary sources such as documents, letters, diaries, official records, and physical artifacts to piece together narratives about what happened, why it happened, and its significance. Unlike scientific experiments, historical events cannot be repeated or tested in highly-structured, controlled environments. Instead, historians critically evaluate the reliability, context, and coherence of sources, often dealing with incomplete or biased records. Interpretation and perspective are crucial in historical analysis, often resulting in varied and sometimes conflicting narratives about the same events among historians.

The overarching goals and objectives of real science and history also differ. Real science seeks to explain natural phenomena and uncover universal laws that apply regardless of time and place. Its primary goal is to generate reliable knowledge that can predict and explain the behavior of the natural and physical world. In contrast, history aims to understand the complexity of human experiences, societal changes, and cultural developments over time. It seeks to provide insights into how past events have shaped the present and to offer lessons that can inform future decisions. History is deeply rooted in the particularities of time and place, emphasizing the uniqueness of individual events and human actions.

Furthermore, the focus of each field reflects its unique aims. Real science often deals with the physical and biological aspects of the universe, exploring everything from the subatomic level to the vastness of space. It encompasses disciplines such as physics, chemistry, biology, and astronomy. History, on the other hand, focuses on human activity, encompassing political, social, economic, and cultural dimensions. It includes subfields like political history, social history, economic history, and cultural history, each examining different aspects of human societies.

In summary, real science and history are distinct yet complementary fields of study. Real science emphasizes empirical evidence and seeks to uncover universal laws through repeatable experiments and quantitative analysis. History focuses on understanding the human past through qualitative analysis and interpretation of historical records. Both fields contribute to a comprehensive understanding of the world around us, with real science explaining the natural phenomena and history providing insights into human experiences and societal development.

CHAPTER 3

The Role of Evidence in Real Science

Evidence plays a central role in real science by serving as the foundation for validating and refining theories and hypotheses (Poincare, 2022). It encompasses observations, experiments, and data that either support or challenge scientific ideas. Empirically-driven evidence is critical for establishing the credibility of scientific claims, ensuring that conclusions are based on valid, reliable, and reproducible results rather than conjecture or bias. By systematically collecting and analyzing empirically derived evidence, scientists develop robust theories, make informed predictions, and enhance both the scientific community's and the public's understanding of natural phenomena, continually testing and refining knowledge in response to new discoveries.

3.1 What is and What isn't Real Science.

Real science is an organized innovativeness that builds capacity and knowledge in the form of testable explanations and predictions about the universe (Chalmers,2013; Patton, 2018; Science, 2024). It relies on the scientific method, a very structured process that involves making observations, formulating hypotheses, conducting experiments, and analyzing data to draw conclusions. Scientific knowledge is distinguished by its foundation in empirically-driven evidence, which means it is based on evidence that can be observed and/or measured. Theories in real science must be falsifiable, meaning they can be tested and potentially disproven. Peer review and replication are also key components, ensuring that findings are carefully examined and re-examined by subject matter experts only later to be validated by the interdisciplinary and multisectoral scientific community.

One of the core principles of real science (Kretser, 2019) is its commitment to objectivity and impartiality. Scientific inquiry seeks to minimize biases (e.g., implicit and/or explicit) and subjectivity by relying on standardized methodology and reproducible results. It is also inherently self-correcting; as new empirical evidence emerges, scientific theories can be revised or replaced. The ever-changing nature of real science enables continuous evolution of scientific theories and deepens our understanding of the natural world. Disciplines like physics, chemistry, biology, and geology exemplify this, offering systematic and verifiable insights into the workings of the universe (Solomon, 2024).

In contrast, what isn't real science can be broadly categorized as pseudoscience (Hansson, 2019), superstition, or other forms of belief that do not adhere to the highly-structured scientific method. Pseudoscience may present itself with scientific terminology and claim legitimacy, but it lacks empirically-driven evidence and does not undergo comprehensive testing or interdisciplinary and multisectoral peer review. Pseudoscience is often characterized by reliance on subjective evidence, resistance to independent testing by other experts, and a lack of progress or adaptation when confronted with contradictory evidence. Examples include astrology (Allum, 2011), reflexology (Edwards-Price, 2021), and certain other pseudoscience theories that are not supported by valid and reliable data or replicable results.

Superstition and belief systems based on faith, tradition, or authority also fall outside the realm of real science (Crossman, 2024). These beliefs are often deeply ingrained in culture and personal experience but do not rely on empirically-driven evidence or the scientific method. While they may offer valuable insights into human behavior and societal values, they do not provide testable or falsifiable explanations of natural phenomena. The distinction between real science and non-real science is essential for maintaining the integrity and reliability of scientific knowledge (Resnik et al., 2023), ensuring that it remains a robust and objective pursuit of truth.

Understanding what is and isn't real science is critical for navigating a world increasingly influenced by scientific and technological advancements. It helps in distinguishing between valid and reliable information and misinformation, fostering critical thinking, logical reasoning, and promoting informed decision-making. By adhering to the core principles of the scientific method, we can better understand the natural world and apply this knowledge in ways that benefit all of society.

3.2 The Principles Underlying Scientific Inquiry

Scientific inquiry is grounded in a set of fundamental principles that guide researchers in their quest to understand natural phenomena (Chang, 2014). These principles ensure that the knowledge acquisition process is organized, efficient, objective, and reproducible, setting scientific inquiry apart from other forms of non-scientific investigation.

One of the foremost principles is empiricism, which emphasizes the role of observation and experimentation in acquiring knowledge (Gooding, 2012). Empirical evidence, derived from sensory experience (i.e., observation) and measurement (i.e., experimentation), forms the foundation of scientific knowledge. Researchers design experiments and observational studies to gather data that can confirm or disprove hypotheses. This reliance on empirical data ensures that scientific claims are based on observable and measurable phenomena rather than on subjective beliefs or unverified assertions.

Falsifiability is another key principle of scientific inquiry, as articulated by philosopher Karl Popper in his book, *The Logic of Scientific Discovery* (1959). For a hypothesis or theory to be considered scientific, it must be testable and capable of being proven false. This principle encourages scientists to design experiments that can potentially disprove their hypotheses, fostering a demanding strategy and a critical approach to research. Falsifiability ensures that scientific theories remain open to revision and refinement in light of new evidence, thereby promoting the continuous advancement of scientific knowledge.

Reproducibility is also a critical principle underlying scientific inquiry (Boylan, 2018). Reproducibility means that experiments and studies should yield consistent results when repeated by different researchers under similar experimental conditions. This principle ensures that scientific findings are valid and reliable and not the result of random chance or experimental error.

Peer review is yet another foundational principle underlying scientific inquiry (Jana, 2019). It involves the evaluation of research by independent subject matter experts in the same discipline, who assess the methodology, data, and conclusions before the work is published. This process helps to maintain the quality and integrity of scientific research by identifying potential errors and ensuring that only strong and credible findings are disseminated.

Objectivity and transparency further support the integrity of scientific inquiry (Nunn et al., 2018). Objectivity requires scientists to strive for impartiality and to minimize personal biases in their research. This involves using standardized methods and clear criteria for data collection and analysis. Transparency requires that scientists openly share their methods, data, and findings with the broader scientific community, enabling others to scrutinize, replicate, and build upon their work. This openness fosters collaboration, accelerates scientific progress, and helps to prevent misconduct and fraud.

Together, these principles form a robust framework for scientific inquiry, ensuring that the pursuit of knowledge is organized, empirically-driven, falsifiable, objective, transparent, reproducible, and continually self-correcting through peer review. By adhering to these principles, real science advances our understanding of the natural world in a reliable and meaningful way, contributing to technological innovations and informed decision-making that benefit all in society.

3.3 The Nature of Scientific Explanations

Scientific explanations aim to provide an understanding of why and how certain phenomena occur in the natural world (Leiss, 2023). These explanations are rooted in the principles of observation, experimentation, critical thinking, and logical reasoning. A key feature of scientific explanations is that they are based on experiential and measurable data, which can be tested and validated through experimentation and observation. By relying on empirically-driven evidence, scientific explanations strive to offer impartial accounts of natural events, minimizing the influence of personal biases or subjective interpretations.

At the heart of scientific explanations are hypotheses and theories (Varpio et al., 2020; Casula, 2021). A hypothesis is a tentative explanation or prediction that can be tested through research and experimentation. If repeated testing and evidence consistently support a hypothesis, it can evolve into a scientific theory. A theory is a well-substantiated explanation of an aspect of the natural world that is supported by a large body of evidence. Unlike everyday use of the term *theory*, in real science, a theory is a highly-structured framework that has withstood extensive testing and scrutiny. Examples include the theory of evolution by natural selection (Sober, 2024) and the theory of general relativity (Plebanski et al., 2024). These theories provide comprehensive explanations for a wide range of phenomena in the natural and physical world and guide further research and discovery.

Scientific explanations also emphasize the importance of causality, aiming to uncover the cause-and-effect relationships underlying natural events (Pearl, 2020). Understanding these relationships allows scientists to predict future occurrences and manipulate variables to observe alternative outcomes. For instance, understanding the causal relationship between microorganisms and disease has led to advancements in medicine and public health (Ristori et al., 2024). The ability to identify and manipulate causal factors is a powerful aspect of scientific explanations, as it enables practical applications and technological innovations.

Another critical aspect of scientific explanations is their openness to revision. Science is inherently self-correcting (Peterson et al., 2021); new evidence can challenge existing explanations, leading to their refinement or replacement. This dynamic process ensures that scientific knowledge remains up-to-date and accurate. For example, the shift from Newtonian mechanics to Einstein's theory of relativity illustrates how scientific explanations evolve with new discoveries and insights (El-Sherbini, 2024). This adaptability is a strength of real science, allowing it to progress and expand our understanding of the natural world.

Moreover, scientific explanations are often framed within broader conceptual models that provide context and coherence (Funari et al., 2021). These models integrate various hypotheses and theories to offer a comprehensive understanding of complex adaptive systems. For example, the model of plate tectonics integrates geological, physical, and chemical evidence to explain the movement of Earth's lithospheric plates (Scotese et al., 2025). Such models are invaluable in synthesizing diverse lines of evidence and guiding further research and discovery.

In summary, scientific explanations are characterized by their reliance on empirical evidence, their focus on causality, their capacity for revision, and their integration within broader conceptual frameworks that provide context and coherence. These features ensure that scientific explanations are testable, reliable, and constantly evolving, providing deeper understandings of the natural world that drive both theoretical progress and practical applications.

3.4 The Role of Experimentation in Real Science

Experimentation plays a crucial role in the scientific process, serving as a primary method for testing hypotheses and generating empirical evidence (Poincare, 2022). Through highly-structured and controlled experiments, scientists can manipulate variables to observe their effects, establishing causal relationships that underpin scientific theories. This systematic manipulation and observation allow researchers to isolate specific factors and understand their direct impact on the phenomena being studied. For instance, in a medical trial, varying the dosage of a drug and observing its effects on patients can provide insights into the drug's efficacy and potential side effects (Rahman et al., 2024).

A key aspect of experimentation is the establishment of controlled conditions, which are essential for minimizing external influences that could skew the results (Antony, 2023). By maintaining consistent conditions except for the variable being tested, researchers can ensure that any observed changes are due to the manipulation of that variable alone. This level of control helps to produce reliable and replicable results, which are fundamental for building and validating scientific knowledge. For example, in physics, experiments conducted in isolated environments, such as vacuum chambers, eliminate peripheral factors that could affect the outcomes (Taurino et al., 2023).

Replication and reproducibility are also integral to the role of experimentation in real science (Nosek et al., 2020; Peng et al., 2021). Once an experiment produces a result, it is crucial that other scientists can replicate the study and achieve similar outcomes under the same conditions. This reproducibility verifies the reliability of the findings and reinforces their validity. If an experiment cannot be replicated, it raises questions about the accuracy and integrity of the initial results. Therefore, detailed documentation of experimental procedures and conditions is crucial (Gatto et al., 2023), enabling other researchers to repeat the experiments and confirm the findings independently.

Furthermore, experimentation drives innovation and discovery (Thomke, 2020). Through experimental research, scientists can explore uncharted territories and test new ideas, often leading to breakthroughs and advancements. For example, the development of advanced medical technologies such as CRISPR (Clustered Regularly Interspaced Short Palindromic Repeats) gene editing tools (Liu et al., 2022) originated from experimental research that explored novel concepts and applications. Experimentation not only tests existing hypotheses but also generates new questions and pathways for exploration, fueling the continuous advancement of real science and our understanding of natural world phenomena (Nanglu et al., 2023).

Experimentation also plays a critical role in both the practical applications and technological development of real science (Liu et al., 2023). In fields such as engineering, medicine, and environmental science, experimental research directly informs the design, development, implementation, and evaluation (i.e., life-cycle of experimentation) of new technologies and interventions (Parolin et al., 2024). By meticulously testing prototypes, treatments, and solutions in controlled settings, scientists and engineers can optimize their effectiveness and safety before widespread application in the public-at-large. This process ensures that new technologies and medical treatments are both efficacious and safe for public use, ultimately contributing to societal progress and improved health outcomes (Subbiah, 2023).

In summary, experimentation is foundational for empirically-driven scientific inquiry, providing a highly-structured and organized method for testing hypotheses, establishing causal relationships, and generating reliable data. With controlled conditions, reproducibility, and meticulous documentation, experimentation upholds the integrity and credibility of scientific findings. It also drives discovery and innovation, leading to practical applications that advance technology and enhance quality of life.

3.5 Criteria for Theory Choice

Choosing between competing scientific theories involves several criteria that ensure the most robust, comprehensive, and reliable explanation is selected (Christie et al., 2023). These criteria are designed to assess the explanatory power, empirical adequacy, and practical utility of theories, guiding scientists toward the most scientifically valid options.

One of the primary criteria for theory choice is empirical adequacy, which refers to how well a theory corresponds with observed data and experimental results (Gooding, 2012). A theory must accurately predict and explain the phenomena within its scope, matching empirical evidence gathered through observation and experimentation. For instance, the theory of general relativity gained widespread acceptance because it could account for anomalies in Mercury's orbit and predict phenomena like gravitational lensing, which were later confirmed through observation (Hartle, 2021). A theory's ability to consistently align with empirical data is a fundamental requirement for its acceptance in the scientific community.

Another critical criterion is coherence, both internally and with existing knowledge (Funari et al., 2021). A theory should be logically consistent, free of contradictions within its own framework. Additionally, it should integrate well with established theories and knowledge across different scientific domains. For example, quantum mechanics (i.e., atomic and sub-atomic levels) and classical mechanics (i.e., macroscopic objects), while seemingly contradictory, both find their place within the broader framework of modern physics, each explaining different scales of physical phenomena (Sakurai et al., 2020). The coherence of a theory with existing knowledge ensures that it contributes to a unified and comprehensive understanding of the natural world (Manoj et al., 2023).

Simplicity, or parsimony, is also a key consideration in theory choice (Falk et al., 2023). Often guided by the principle known as Occam's Razor (Gross, 2019), simplicity favors theories that make fewer assumptions and offer more straightforward explanations without unnecessary complexity. While simplicity alone is not a definitive criterion, a simpler theory that equally explains the phenomena is preferred over a more complex one. For instance, heliocentrism, which posits that the Earth and other planets orbit the Sun, eventually replaced the geocentric model because it provided a simpler and more accurate explanation of planetary motions without complex epicycles (Gaida, 2022).

Explanatory power and scope are additional important criteria for choosing between theories (Jansson, 2021). A strong scientific theory should not only explain known phenomena but also predict new, previously unobserved phenomena. The broader the range of phenomena a theory can explain and predict, the more powerful and valuable it is considered. The theory of evolution by natural selection (Avery, 2021), for instance, provides a comprehensive framework that explains a vast array of biological observations and has predicted discoveries in genetics, paleontology, and ecology.

Fruitfulness is another valuable choice criterion, referring to a theory's ability to generate new research questions, hypotheses, and areas of investigation (Cozzo, 2023). A fruitful theory opens up new avenues for discovery and experimentation, driving scientific progress. Watson and Cricks' 1953 discovery of DNA's double helix structure, guided by the principles of molecular biology (Allison, 2021; Shen, 2023), has led to numerous advancements in genetics, biotechnology, and medicine, exemplifying how a fruitful theory can greatly impact multiple scientific disciplines.

Finally, the pragmatic choice criterion of utility considers the practical applications of a theory (Pezzullo et al., 2023). Theories that lead to technological advancements, medical breakthroughs, or improved experimental methodologies in various scientific fields are highly valued (Oschman, 2015). For example, the development of semiconductor theory has been instrumental in the creation of modern electronics, demonstrating the practical benefits of a well-founded scientific theory (Razavi, 2021).

In summary, the criteria for theory choice include empirical adequacy, coherence, simplicity, explanatory power, fruitfulness, and utility. These criteria collectively ensure that the chosen theory provides a robust, comprehensive, and practical explanation of natural phenomena, contributing to the advancement of scientific knowledge and its applications. By rigorously evaluating theories against these standards, scientists can navigate complex questions and build a more accurate understanding of the world.

3.6 Types of Evidence: Empirical, Subjective, and Theoretical

Empirical evidence is the cornerstone of scientific inquiry, derived from systematic observation, experimentation, and data collection (Coccia, 2018). This type of evidence is based on direct, measurable, and observable phenomena, making it the most reliable for validating scientific theories and hypotheses. For example, in a clinical trial for a new medication (Huang et al., 2020), empirical evidence is gathered through controlled experiments and statistical analysis of patient outcomes. This evidence provides a concrete basis for drawing conclusions and making predictions about the efficacy and safety of treatments. Empirical evidence is crucial because it can be independently verified and reproduced, ensuring that scientific findings are robust and objective (Adler, 2022).

Subjective evidence, in contrast, is based on personal accounts, observations, or individual experiences rather than systematic research (Smedberg, 2021). This type of evidence is often considered less reliable in scientific contexts because it lacks the rigor and control of empirical methodology. While subjective evidence can generate hypotheses or highlight areas for further research, it is not sufficient on its own to support scientific claims due to its susceptibility to bias and lack of generalizability. For instance, testimonials about a particular diet's effectiveness might offer valuable personal insights, but they do not provide the comprehensive data needed to establish a scientifically valid conclusion (Ruxton, 2022).

Theoretical evidence involves the use of models, simulations, and theoretical frameworks to explain phenomena and predict outcomes (Baum et al., 2024). This type of evidence is derived from the logical and mathematical analysis of scientific theories, often based on established principles and laws. For example, theoretical evidence in physics might involve complex mathematical models to predict the behavior of particles under certain conditions (Ramstead et al., 2023). While theoretical evidence is essential for developing and testing scientific ideas, it relies on the assumptions and accuracy of the underlying models. It is often validated by comparing theoretical predictions with empirical data. In this way, theoretical evidence complements empirically-driven evidence, providing a deeper understanding of the mechanisms behind observed phenomena (Cornelissen, 2023).

In summary, while empirical evidence is the gold standard for validating scientific claims due to its direct, observable nature, subjective evidence can offer initial insights and personal perspectives but lacks scientific rigor. Theoretical evidence plays a critical role in developing and explaining scientific concepts, but its validity is ultimately tested against empirical observations. Each type of evidence contributes uniquely to the scientific process, helping to build and refine our understanding of the natural world.

3.7 The Importance of Reproducibility and Peer Review.

Reproducibility is a fundamental principle in real science that ensures the reliability and validity of research findings (Buzbas, 2023). It refers to the ability of different researchers to replicate the results of a study using the same methods and conditions. The importance of reproducibility lies in its role as a safeguard against errors and biases, providing confidence that findings are not the result of chance or methodological flaws. When a study's results can be reproduced, it confirms the robustness of the conclusions and strengthens the evidence supporting a particular theory or hypothesis. Reproducibility also allows scientists to build upon previous work, advancing knowledge through a cumulative and reliable foundation of verified findings.

Peer review is another crucial component of the scientific process that contributes to the credibility and quality of research (Willis, 2024). It involves the evaluation of a study by independent experts in the same field before it is published in a scientific journal. The peer review process helps identify errors, biases, and methodological issues that the original researchers might have overlooked, ensuring that only high-quality, rigorous research is disseminated. By providing constructive feedback and recommendations for improvement, peer reviewers help enhance the clarity, accuracy, and significance of scientific work. Peer review also serves as a form of quality control, maintaining the integrity of the scientific literature and fostering trust in the research community.

Together, reproducibility and peer review are essential for maintaining the standards of scientific inquiry. Reproducibility ensures that research findings are consistent and reliable, while peer review provides a critical assessment of the study's validity and impact. Both processes contribute to the cumulative advancement of knowledge by ensuring that scientific claims are well-founded and subject to rigorous scrutiny. This collaborative and self-correcting nature of real science helps to build a robust and trustworthy body of evidence, driving progress and innovation in various fields of study.

3.8 The Challenge of Distinguishing Between Correlation and Causation.

Differentiating between correlation and causation presents a major challenge in scientific research (Anjum et al., 2024), as failing to do so can result in incorrect conclusions and the development of flawed policies and methodologies.

Correlation refers to a statistical relationship between two variables, where changes in one variable are associated with changes in another. For example, a correlation might be observed between ice cream sales and drowning incidents, suggesting that as ice cream sales increase, so do drowning incidents. However, this does not imply that eating ice cream causes drowning; rather, both are related to a third variable, such as hot weather, which influences both variables independently. Correlation alone does not establish a cause-and-effect relationship, making it necessary to look beyond mere associations to understand underlying mechanisms (Zemla et al., 2023).

Causation involves a direct cause-and-effect relationship where one variable directly influences another. To establish causation, researchers must demonstrate that changes in one variable led to changes in another and that this relationship is not due to confounding factors. Experimental studies, such as randomized controlled trials, are often used to establish causation by manipulating one variable and observing the effect on another while controlling for other variables (Stoner et al., 2023). For example, a clinical trial testing a new medication involves assigning participants to treatment or placebo groups to observe the effects of the medication, thereby establishing a causal link between the medication and health outcomes (Rahman et al., 2024).

The challenge in distinguishing between correlation and causation arises from the need to control for confounding variables that might influence the observed relationship (Chu et al., 2023). Without proper experimental design or statistical controls, it is easy to mistakenly interpret a correlation as causation. For instance, a study might find a correlation between increased screen time and poor academic performance (Oswald et al., 2020), but without considering other factors like socioeconomic status or parental involvement, the causal relationship remains unclear. Researchers must use a combination of experimental methods, statistical techniques, and theoretical frameworks to identify and isolate causal relationships accurately (Huntington-Klein, 2021)

Overall, while correlation can provide valuable insights and suggest potential areas of investigation, establishing causation requires rigorous research methods and careful consideration of confounding variables. By addressing these challenges, scientists can better understand the true nature of relationships between variables and make more informed conclusions that advance empirically-driven knowledge and inform decision-making.

CHAPTER 4

The Evolution of Scientific Theories.

The evolution of scientific theory is a dynamic process driven by ongoing observation, experimentation, and refinement. Initially, a scientific theory emerges as a hypothesis, offering a preliminary explanation for observed phenomena. As evidence is gathered through research and experimentation, the theory is tested and modified to better fit new data. Over time, theories are refined or replaced as new discoveries challenge existing explanations and provide deeper insights. This iterative process ensures that scientific theories remain robust and accurate, reflecting the best available understanding while adapting to new evidence and perspectives. Through this continuous evolution, real science advances and enhances our grasp of the natural world (Fortunato et al., 2018).

4.1 Role of History and Philosophy in Real Science.

The history and philosophy of real science (HPS) is a multidisciplinary field that explores the development and conceptual foundations of science (Verburgt, 2024). By examining the evolution of scientific ideas, practices, and institutions, HPS provides insights into how scientific knowledge is constructed and validated (Pradeu et al., 2024). This field not only traces the chronological development of scientific theories and discoveries but also delves into the sociocultural and intellectual contexts that shape scientific endeavors (Laudan et al., 2020). Through historical analysis, HPS can reveal the contingent and sometimes non-linear nature of scientific progress, challenging the notion of real science as a straightforward accumulation of facts (Hooker et al., 2022).

Philosophy of real science, on the other hand, addresses foundational questions about the nature and methodology of real science (Rosenberg et al., 2019). It interrogates the principles underlying scientific inquiry, such as the nature of scientific explanations, the role of experimentation, and the criteria for theory choice (Bailer-Jones, 2009). Philosophers of real science debate issues like realism versus anti-realism (Brock et al., 2014), the demarcation problem (i.e., what distinguishes real science from pseudoscience) (Resnik et al., 2023), and the implications of scientific revolutions (Kuhn, 1962). These discussions are crucial for understanding the epistemic value (e.g., a kind of value which attaches to cognitive successes such as true beliefs, justified beliefs, knowledge, and understanding) and metaphysical assumptions (e.g., fundamental beliefs about the nature of reality, existence, and the universe that are often not subject to empirical verification and are foundational to various philosophical and religious systems) that underpin scientific theories (Psillos, 2008) and for evaluating the objectivity and reliability of scientific knowledge (Alai, 2017).

Together, the history and philosophy of real science foster a critical awareness of how scientific knowledge is produced and validated (Schikore, 2011). They encourage scientists, scholars, and the public-at-large to reflect on the broader implications of scientific practices and discoveries (Vos, 2021). This reflection is essential for addressing contemporary issues in real science and technology, such as ethical considerations in research, the societal impact of scientific advancements, and the communication of scientific knowledge (Chakravartty, 2023). By integrating historical and philosophical perspectives, HPS offers a richer and more nuanced understanding of real science as a dynamic and human endeavor (Arabattzis et al., 2012).

4.2 The Rhetorical Nature of Real Science.

The rhetorical nature of real science highlights how scientific communication is not merely about presenting objective facts, but also involves persuasive elements aimed at convincing various audiences of the validity and significance of scientific claims (Fahnestock, 2019). Scientists use rhetorical strategies (Peeples et al., 2022) to frame their findings, argue for the acceptance of their theories, and garner support from their peers, funding bodies, and the public. This perspective underscores that scientific papers, presentations, and debates are crafted with careful attention to language, structure, and audience engagement, employing techniques that enhance credibility, clarity, and appeal (Seethaler et al., 2024).

One aspect of the rhetorical nature of real science is the use of ethos, pathos, and logos (Slater, 2023). *Ethos* refers to the credibility of the scientist, which is often established through credentials, affiliation with reputable institutions, and a history of prior research accepted by the scientific community. *Pathos* involves appealing to the audience's emotions, which can be subtle in scientific contexts but might include emphasizing the societal benefits or urgent implications of research findings. *Logos*, the logical structure of the argument, is central in real science, where data, evidence, and logical reasoning must be meticulously presented to demonstrate the validity of conclusions. Together, these rhetorical elements help scientists construct persuasive arguments and compelling narratives.

Additionally, the rhetorical nature of real science is evident in how scientific controversies and major paradigm shifts are negotiated (Caudill, 2023). When new theories challenge established paradigms, proponents must not only present robust empirical evidence but also persuade the scientific community to reconsider entrenched understandings. Historical examples, such as the acceptance of heliocentrism (Ng, 2023) or the theory of evolution by natural selection (Papale et al., 2024), illustrate how rhetoric plays a crucial role in facilitating scientific change. Scientists advocating for new ideas must effectively communicate the shortcomings of existing models and the advantages of their proposals, often engaging in debates and discussions that are as much about persuasion as they are about evidence.

Understanding the rhetorical nature of real science also sheds light on the public communication of real science (Bonyadi et al., 2024). Scientists and science communicators must translate complex concepts into accessible language without oversimplifying or misrepresenting the real science. This involves using rhetorical tools to make real science engaging and relevant to diverse audiences, fostering public understanding and trust in scientific processes (Bromme et al., 2024). In an era where misinformation can spread rapidly (DeVerna et al., 2024), the ability of scientists to persuasively and accurately communicate their work is more important than ever for ensuring informed public discourse and policy-making (Smillie et al., 2024).

4.3 Insight on the Human Aspect of Real Science

The human aspect of real science emphasizes that real science is a profoundly human endeavor, shaped by the values, motivations, and social contexts of the individuals and communities involved in scientific activities (Oruh et al., 2024). Scientists, like all people, are influenced by their backgrounds, experiences, and personal biases, which can affect the questions they ask, the methods they use, and the interpretations they draw from their data. Recognizing the human element in real science helps to understand that scientific knowledge is not produced in a vacuum but within a complex web of social interactions and cultural norms.

One significant dimension of the human aspect of real science is collaboration and teamwork (Al Hamad et al., 2024). Scientific research often involves multidisciplinary teams working together, combining their subject matter expertise to tackle complex problems. This collaborative nature of real science fosters the exchange of ideas, promotes innovation, and accelerates progress. However, it also requires effective communication, mutual respect, and the ability to navigate interpersonal dynamics. The human aspect of collaboration highlights the importance of leadership, mentorship, and the cultivation of an inclusive and supportive research environment where diverse perspectives are valued.

Furthermore, the human aspect of real science acknowledges the ethical and moral responsibilities of scientists (Stewart, 2023). Decisions about research directions, experimental practices, and the application of scientific findings have profound implications for society and the environment. Scientists must consider the potential impacts of their work and strive to conduct research that benefits humanity while minimizing harm. Ethical considerations, such as informed consent in human studies, animal welfare in experimental research, and the responsible use of emerging advanced scientific technologies, are fundamental to the practice of real science and reflect the broader societal values in which real science operates.

Lastly, the human aspect of real science involves the personal journeys and stories of scientists themselves (Rogers et al., 2023). The careers of scientists are often marked by passion, curiosity, perseverance, and sometimes significant personal and professional challenges. These narratives humanize real science, making it more relatable and inspiring to the public and future generations of scientists. Celebrating the achievements and addressing the struggles of scientists from diverse backgrounds can also help to promote inclusivity and equity in the scientific community, ensuring that talent and innovation are nurtured across all sectors of society.

4.4 The Growth of Scientific Knowledge

The growth of scientific knowledge is a dynamic and cumulative process that involves the continuous refinement, expansion, and sometimes, revision of existing understandings (Di Ciccio et al., 2015). This growth is driven by systematic observation, experimentation, and the development of theories that explain natural phenomena. Over time, incremental advancements build upon previous discoveries, creating a more comprehensive and sophisticated body of knowledge. However, scientific progress is not always linear; it often involves periods of rapid innovation, known as scientific revolutions, where paradigms shift, and fundamentally new frameworks emerge (Kuhn, 1962).

Central to the growth of scientific knowledge is the rigorous application of the scientific method (Rouse, 2018). As previously discussed, this involves formulating hypotheses, conducting experiments to test these hypotheses, and analyzing the results to draw conclusions. Peer review and replication of results are crucial components of this process, as they ensure that findings are scrutinized and validated by the scientific community. This collective effort helps to minimize errors and biases, contributing to a more reliable and robust body of knowledge. As new tools and technologies become available, they enable scientists to explore previously inaccessible areas, leading to breakthroughs that further accelerate the growth of knowledge.

Collaboration and interdisciplinary research also play a significant role in the expansion of scientific knowledge (Hillersdal et al., 2023). By bringing together experts from different fields, interdisciplinary approaches can address complex problems that single disciplines might not be able to solve alone. For example, the intersection of biology, chemistry, and computer science has given rise to bioinformatics (Gauthier et al., 2019), a field that has revolutionized our understanding of genetic information and disease mechanisms at the molecular and sub-molecular level of organization. Such collaborations often lead to innovative methods and new lines of inquiry, demonstrating that the growth of scientific knowledge is often fueled by the convergence of diverse perspectives and expertise.

The advancement of scientific knowledge is shaped by social, economic, and political factors (Biermann et al., 2022). Funding priorities, policy decisions, and public interest influence the direction and focus of research. During periods of societal need, such as public health crises or environmental challenges, scientific progress often accelerates to meet urgent demands, as seen with the rapid development of RNA vaccines during the COVID-19 pandemic. Conversely, political or economic instability can hinder scientific advancements. Acknowledging these influences underscores the need to cultivate a stable, supportive environment for research and to foster a culture that prioritizes and invests in scientific inquiry.

The advancement of scientific knowledge embodies humanity's enduring pursuit to understand the natural world and enhance the well-being of all living things. It stands as a testament to real science, which fuels the curiosity, ingenuity, and collaborative spirit of both individual scientists and the broader scientific community. By continually pushing the boundaries of knowledge, real science not only uncovers new insights but also raises new questions, ensuring that the pursuit of understanding remains an ever-evolving endeavor. This journey must be carefully structured, guided by systematic processes that build sustainable capacity and lay the foundation for future generations of scientists.

4.5 How Scientific Theories Develop and Evolve Over Time.

Theories in real science develop and evolve over time through a process of observation, experimentation, and refinement (Larson, 2006). Initially, a theory often begins as a hypothesis—a tentative explanation for a particular phenomenon based on limited evidence. Scientists test this hypothesis through experimentation and observations, collecting data that either supports or challenges it. As empirical evidence accumulates, the hypothesis may be refined, expanded, or even discarded in favor of a more accurate explanation. For example, the early models of atomic structure, such as Dalton's solid sphere model, evolved through the work of scientists like Thomson, Rutherford, and Bohr as new experimental data became available (Omilani, 2024). Each new model built upon or corrected the previous one, leading to our current understanding of the atom.

Over time, as a theory gains empirical support, it becomes widely accepted within the scientific community. However, acceptance is not the end of a theory's development. Theories are continually tested and challenged (Sandberg et al., 2021), particularly as new technologies and methods allow for more precise measurements or observations. This ongoing scrutiny is a fundamental aspect of the scientific method, ensuring that theories remain robust and reflective of the best available evidence. For instance, Einstein's theory of general relativity (Plebanski et al., 2024) which revolutionized our understanding of gravity, continues to be tested more than a century after its formulation, with new observations, such as those from gravitational wave detectors, offering further confirmation and insights.

The evolution of theories is also influenced by paradigm shifts—moments when new evidence or perspectives radically change the foundational assumptions of a scientific field (Kuhn, 1962). Such shifts often occur when existing theories can no longer adequately explain observed phenomena, leading to the development of new theories that better account for the data. A classic example is the shift from the geocentric model of the universe, which placed the Earth at the center, to the heliocentric model proposed by Copernicus, which correctly identified the Sun as the center of the solar system (El-Sherbini, 2024). This shift fundamentally and completely altered the way humanity understood its place in the cosmos and paved the way for modern astronomy.

In addition to paradigm shifts, interdisciplinary and multisectoral research can lead to the evolution of theories by introducing new ideas and methods from different fields (Newman, 2024). As scientists from diverse disciplines and sectors collaborate, they bring different perspectives that can challenge existing theories and inspire creative and innovative approaches to old problems. For instance, the intersection of biology and information technology has given rise to bioinformatics (Buttar et al., 2024), a field that has transformed our understanding of genetics by applying computational theories and methodologies to biological data. This cross-pollination of ideas and knowledge accelerates the evolution of theories, often leading to breakthroughs that would not have been possible within the confines of a single siloed discipline.

Ultimately, the development and evolution of theories are a testament to the self-correcting nature of science (Donley, 2024). Theories are not static; they are dynamic constructs that adapt to new evidence, knowledge, and ideas. This adaptability ensures that scientific knowledge continues to grow and improve over time, leading to a deeper and more accurate understanding of the natural world. As new scientific discoveries are made and new challenges arise, theories will continue to evolve, reflecting the ever-changing landscape of scientific inquiry.

4.6 Paradigm Shifts: From Newtonian Physics to Quantum Mechanics

Paradigm shifts in science mark profound and radical changes in the fundamental understanding of natural phenomena (Kuhn, 1962), and one of the most notable examples is the transition from Newtonian physics to Quantum Mechanics (Zuccarini et al. 2024; Fliessbach, 2024). Newtonian physics, established by Sir Isaac Newton in the 17th century, provided a comprehensive framework for understanding classical mechanics. Newton's laws of motion and universal gravitation successfully described the behavior of macroscopic objects and celestial bodies, forming the foundation of classical physics. This paradigm was remarkably successful in explaining a wide range of physical phenomena, from the orbits of planets to the trajectories of projectiles.

However, as the global scientific community delved into the behavior of particles at the atomic and subatomic levels, they encountered phenomena that could not be explained by Newtonian mechanics. In the early 20th century, experiments revealed that particles such as electrons exhibited both wave-like and particle-like properties (Fleury, 2024), a duality that challenged the deterministic nature of classical physics. This led to the development of quantum mechanics, a revolutionary paradigm shift that introduced radically-new principles in physics to explain the behavior of matter and energy at microscopic scales. Quantum mechanics (Scherrer, 2024), developed through the groundbreaking contributions of esteemed physicists such as Max Planck, Albert Einstein, Niels Bohr, and Werner Heisenberg, explains phenomena like energy quantization, the uncertainty principle, and wave-particle duality—concepts that fundamentally diverge from the principles of Newtonian physics.

The transition from Newtonian physics to Quantum Mechanics was not simply an incremental evolution but a fundamental revolution in both scientific thought and public understanding of the universe. Quantum mechanics introduced a probabilistic framework in which the precise position and momentum of atomic and subatomic particles cannot be simultaneously determined with absolute certainty—a sharp departure from the deterministic predictions of Newtonian mechanics. This paradigm shift redefined our perception of reality, demonstrating that at the quantum level, classical concepts of causality and predictability give way to probabilities and inherent uncertainties.

Despite its fundamental departure from classical physics, Quantum Mechanics has demonstrated remarkable success in explaining and predicting phenomena at the microscopic scale (Wang et al., 2023; Nałęcz-Charkiewicz et al., 2024). Its principles have driven numerous technological advancements, including semiconductors, lasers, and MRI machines, which are now essential to modern life. The scientific community's acceptance of Quantum Mechanics did not invalidate Newtonian physics but rather expanded the boundaries of scientific understanding. While Newtonian mechanics remains a highly effective approximation for macroscopic objects and everyday experiences, Quantum Mechanics provides the essential framework for describing the behavior of particles at the atomic and subatomic levels.

This profound paradigm shift exemplifies how scientific revolutions can reshape our understanding of the natural world (Hanna, 2024). The transition from Newtonian physics to Quantum

Mechanics not only resolved inconsistencies at the atomic and sub-atomic scales but also demonstrated the adaptability of real science in the face of new evidence. It underscores the dynamic nature of scientific inquiry, where groundbreaking ideas emerge to challenge the limitations of established theories, driving further discoveries that transform both the scientific community and public perception (Grinin et al., 2024).

4.7 The Impact of Technological Advances on Scientific Discovery.

Technological advances have profoundly impacted scientific discovery, transforming how research is conducted and expanding the frontiers of knowledge (Volti et al., 2024). Innovations in technology often enable scientists to observe and measure phenomena with unprecedented precision, leading to breakthroughs that were once previously unattainable. For example, the development of advanced telescopes (McElwain et al., 2023), such as the Hubble Space Telescope, has allowed astronomers to explore distant galaxies and gain insights into the universe's structure and evolution. Similarly, high-resolution imaging technologies, like electron microscopes, have revealed the intricate details of cellular, molecular, and sub-molecular structures, deepened our understanding of biological morphology and processes and facilitated advancements in fields such as genetics, cell and molecular biology, and medicine (Pattison et al., 2024).

Furthermore, technology has revolutionized data collection and analysis, accelerating the pace of scientific discovery (Kahn et al., 2024). Computational tools and software have enabled researchers to handle vast amounts of data, perform complex simulations, and model intricate systems with greater accuracy. For instance, in climate science, sophisticated climate models powered by high-performance computing (Govett et al., 2024) have improved our ability to predict weather patterns, assess climate change impacts, and develop strategies for mitigation and adaptation. The advent of big data analytics has similarly transformed fields like genomics (Lee, 2024) and public health science (Adenayi et al., 2024), where massive datasets are used to identify patterns, correlations, and causal relationships that inform public health decisions and personalized medicine.

Technological advances also facilitate interdisciplinary and multisectoral research by providing tools and platforms that bridge different scientific fields (Li et al, 2024). For example, advancements in molecular biology and bioinformatics have fostered the development of systems biology (Saranya et al., 2024), an interdisciplinary approach that integrates biology, computer science, and mathematics to understand complex biological systems. This cross-disciplinary collaboration has led to significant discoveries in areas such as gene regulation, metabolic pathways, and drug development. Moreover, the rise of collaborative online platforms and communication technologies (Olaniyi et al., 2024) has enabled researchers from around the world to share knowledge, resources, and findings more effectively, driving innovation and accelerating scientific progress on a global scale.

In addition, technology has democratized access to scientific tools and resources (Galdames et al., 2024), allowing researchers from diverse backgrounds and institutions to contribute to discovery. Open-access journals, online databases, and cloud computing platforms have made it easier for scientists to publish their work, access research data, and collaborate across borders (Khang et al., 2024). This increased accessibility fosters a more inclusive scientific community and accelerates the dissemination of knowledge, enabling discoveries to be built upon and refined more rapidly.

Overall, technological advances have not only enhanced the precision, scope, and scale of scientific research but also fundamentally transformed how scientific discoveries are made, shared, and applied. As technology continues to evolve over time, it will undoubtedly drive further innovations and discoveries, continuing to shape and expand our understanding of the natural world and its complexities.

PART II

The Challenges in Real Science

Real science confronts numerous challenges, including ethical dilemmas, funding limitations, and the accelerating pace of technological advancement. The increasing complexity of global issues—such as institutional bias, racism, climate change, natural disasters, pandemics, economic inequalities, and biodiversity loss—presents an existential threat to a world inhabited by over 8 billion people (Undheim, 2024).

Addressing these issues requires interdisciplinary and multisectoral approaches, but collaboration within and across scientific disciplines is often hindered by differences in methodologies and jargons. Additionally, the pressure to publish in high-impact journals can lead to a publish or perish culture (Becker et al., 2023), sometimes compromising the integrity of the research process. Balancing innovation with ethical responsibility (Teymourifar, 2024), ensuring equitable access to advancements within the scientific community (Shelton et al., 2024), and maintaining public trust in science (Rosman et al., 2024) are ongoing struggles in this dynamic and evolving field of real science.

CHAPTER 5

Pseudoscience and Misconceptions

Pseudoscience and misconceptions pose significant challenges and barriers to public understanding and trust in real science (Danielson et al., 2024; Thompson et al., 2024). Pseudoscience often presents itself with the appearance of legitimacy, using scientific jargon and cherry-picked data, but lacks the rigorous methodology and peer review that characterize true scientific inquiry in the scientific community. Misconceptions, often driven by the deliberate spread of misinformation or the unintentional actions of well-meaning but misinformed individuals, can rapidly gain traction—especially through social media. These false beliefs can result in harmful consequences, such as vaccine hesitancy (Prasad et al., 2024) or reliance on unproven medical treatments (Amir-Azodi et al., 2024). Addressing pseudoscience requires strong science communication, education in critical thinking, and a steadfast commitment to promoting evidence-based knowledge (Lynn et al., 2024).

5.1 Characteristics of Pseudoscience: What It is and How to Identify It.

5.1.1 What is Pseudoscience?

Pseudoscience refers to a set of beliefs, theories, or practices that claim to be scientific but lack the empirical evidence, methodological rigor, and reproducibility that are fundamental to authenticated and real science comprehension (Hansson, 2024). Unlike genuine scientific inquiry, which is grounded in the scientific method and relies on peer review and rigorous, highly-structured and organized testing, pseudoscience often relies on subjective evidence, untestable claims, and selective data and information. Pseudoscience often mimics the appearance of real science, employing technical language and referencing scientific concepts, yet it fails to uphold the rigorous principles that make science a reliable method for understanding the natural world. Examples of pseudoscientific disciplines include astrology, reflexology, and certain alternative and functional medical practices that lack scientific validation (Souza et al., 2024).

5.1.2 Characteristics of Pseudoscience and How to Identify It.

Pseudoscience can be recognized by several defining characteristics (Hansson, 2024). One of its primary traits is its reliance on anecdotal evidence, personal testimonials, or unverified claims rather than controlled, reproducible experiments that withstand rigorous scientific scrutiny. Unlike real science, which is grounded in empirical data and systematic methodology, pseudoscience often makes bold assertions that are not testable or falsifiable, meaning they cannot be disproven through objective investigation.

A hallmark of pseudoscience is its lack of peer review or acceptance within the broader scientific community. Instead of engaging in the rigorous process of scientific validation, proponents frequently appeal directly to the public, positioning themselves as mavericks or rebels challenging the so-called "scientific establishment." This tactic fosters distrust in legitimate scientific discourse while elevating unverified or debunked theories.

Moreover, pseudoscience tends to resist change, holding onto outdated or discredited ideas rather than evolving in response to new evidence. Real science, by contrast, embraces revision and refinement as new discoveries emerge. Additionally, pseudoscientific claims are often marked by confirmation bias, selectively presenting information that supports a predetermined conclusion while disregarding or dismissing contradictory data.

To effectively identify pseudoscience, it is essential to remain vigilant for these red flags. Critically evaluating whether claims are supported by robust, peer-reviewed research and determining if the methods align with the principles of the scientific method can help differentiate genuine scientific inquiry from misleading pseudoscientific rhetoric.

5.1.3 Comparing Real Science to Pseudoscience.

Real science and pseudoscience differ fundamentally in their approach to understanding the natural world, though they may sometimes appear superficially similar. Real science is built upon the scientific method, a structured and systematic process that involves forming hypotheses, designing controlled experiments, collecting and analyzing data, and drawing conclusions based on empirical evidence. This process is transparent, repeatable, and subject to rigorous peer review by subject matter experts. Scientific theories are not static; they are continually tested, refined, or even discarded in light of new evidence, ensuring that knowledge evolves over time. Real science thrives on skepticism, encouraging critical questioning and continuous inquiry to expand our understanding through objective, data-driven investigation.

In contrast, pseudoscience lacks this methodological rigor and structure. Instead of following an open-ended process of discovery, it often begins with a predetermined conclusion and selectively gathers evidence to support preexisting beliefs while ignoring or dismissing contradictory data. Pseudoscientific claims are frequently not testable, and even when they are, the methods used tend to be flawed, biased, or unrepeatable. Unlike real science, which relies on empirical data, pseudoscience often appeals to emotions, subjective experiences, or authority figures rather than objective, systematically-derived evidence. Furthermore, while real science is self-correcting and adapts in response to new findings, pseudoscience is resistant to change, clinging to discredited ideas even in the face of overwhelming counterevidence.

The consequences of these differences are profound. Real science has driven technological innovations, medical advancements, and a deeper understanding of the universe, leading to tangible improvements in human life. Pseudoscience, however, can perpetuate misinformation, hinder scientific progress, and in some cases, cause direct harm—such as when unproven medical treatments are promoted over evidence-based medical care, leading to negative health outcomes (Brinsfield et al., 2024). Distinguishing between real science and pseudoscience requires a critical examination of the methodology behind claims, the quality of the evidence presented, and the openness of proponents to engage with opposing viewpoints and new information. In an era of rapid information dissemination, fostering scientific literacy, critical thinking, and logical reasoning skills is essential to ensuring that knowledge is built on a foundation of rigorous inquiry rather than misleading rhetoric.

5.2 Common Misconceptions in Real Science.

Misconceptions about real science are widespread and can stem from misunderstandings, oversimplifications, or the deliberate spread of misinformation. One common misconception is the belief that scientific theories are just "guesses" or unproven ideas (Tabor et al., 2015). In reality, a scientific theory is a well-substantiated explanation of some aspect of the natural world, supported by a body of evidence and repeatedly tested through observation and experimentation. Theories like evolution and climate change are often misrepresented as being uncertain or debatable, even though they are supported by extensive research and are fundamental to our understanding of systems biology and environmental science.

Another prevalent misconception is the idea that real science is infallible or that scientists are always in agreement (Ziman et al., 1991). Real science is an ongoing process of discovery, and while it is based on empirically-derived evidence, it is also subject to revision as new data emerges. This is sometimes misunderstood as a weakness, leading to the erroneous belief that if scientists change their minds, the science is unreliable. However, this adaptability is a strength, as it allows real science to self-correct and improve over time.

Misinterpretations of scientific uncertainty can also lead to the mistaken belief that if real science cannot provide absolute answers, it is not truthful and trustworthy. In reality, uncertainty is a natural part of scientific inquiry, and understanding the degrees of confidence in scientific findings is crucial for informed scientific decision-making (Develaki, 2024).

Additionally, there is a misconception that correlation implies causation (Anjum et al., 2024b). This fallacy occurs when people assume that because two events occur together, one must cause the other. In reality, many factors could contribute to an observed correlation, and rigorous testing is required to establish a causal relationship. This misunderstanding is often exploited in pseudoscience, leading to misleading claims about the effectiveness of products (Betts et al., 2024) or the dangers of certain behaviors (Xiao, 2024).

Finally, the myth of the "lone genius" persists in popular culture, where scientific breakthroughs are often attributed to individual brilliance of the scientist rather than collaborative effort (Anjum et al., 2024c). While individual scientists have made significant contributions, real science is largely a collective endeavor, requiring the collaboration of diverse interdisciplinary and multisectoral teams, subject matter expert peer review, and the cumulative work of many researchers over time. This misconception can obscure the importance of teamwork and the incremental and iterative nature of scientific progress.

5.3 The Consequences of Pseudoscience in Public Policy and Health.

5.3.1 Consequences of Pseudoscience in Public Policy.

Pseudoscience can have serious and far-reaching consequences when it influences public policy (Pasternack- Taschner et al., 2024). When decision-makers rely on pseudoscientific ideas, they may design, develop, and implement public policies that are not grounded in empirically-driven evidence, leading to ineffective or even harmful outcomes. For example, public policies that deny or downplay the reality of Anthropocene climate change can delay critical actions needed to mitigate its effects (Elabbar, 2024), exacerbating environmental degradation and increasing the risk of natural disasters. Pseudoscience can also fuel public distrust in legitimate scientific research (Lupia et al., 2024; Post et al., 2024), making it more difficult to gain support for public policies that address pressing issues like essential public health services and function, STEM education, and environmental health and protection. When pseudoscientific claims shape public policy, resources may be diverted from effective solutions to unproven or discredited approaches, ultimately putting society in harm's way (Moser, 2024; Lipinski et al., 2024), observed during the COVID-19 pandemic (e.g., misinformation about safe distancing and masking as major public health practices to avoid COVID infection).

5.3.2 Consequences of Pseudoscience in Health.

The impact of pseudoscience on health can be particularly dangerous, often leading to misguided medical decisions, delayed treatments, and preventable deaths. When individuals or communities embrace pseudoscientific beliefs, they may reject proven medical interventions in favor of unproven, ineffective, or even harmful alternatives

(Schulz et al., 2024). This not only jeopardizes individual health but can also contribute to larger public health crises.

One of the most well-documented examples is the anti-vaccine movement (Daubs, 2024), which is rooted in debunked pseudoscientific claims that vaccines cause autism in children. Despite overwhelming scientific evidence disproving this myth, vaccine hesitancy has led to the resurgence of preventable communicable diseases like measles, pertussis, and polio—diseases that had been largely controlled or eradicated in many parts of the world. The refusal to vaccinate not only endangers those who remain unprotected but also compromises herd immunity, putting immunocompromised individuals and those unable to receive vaccines at greater risk.

Beyond vaccines, pseudoscience in health promotes unregulated supplements, alternative therapies, and so-called "miracle cures" that lack scientific validation (Jafari, 2024). From unproven herbal remedies to extreme dietary regimens, these products often make bold claims without any credible clinical evidence. In some cases, they may directly harm patients by interfering with legitimate treatments, causing toxicity, or fostering a false sense of security that delays necessary medical care. The financial cost is also significant, as individuals invest in ineffective treatments while potentially neglecting proven medical interventions.

The dangers of pseudoscience extend beyond individual health choices to broader public health policies. When misinformation shapes decision-making at the governmental or institutional level, it can undermine trust in healthcare delivery systems and clinicians (Bromme et al., 2024; Cullinan et al., 2024). This erosion of trust creates barriers to accessing evidence-based medical care, particularly for vulnerable and marginalized populations who may already face disparities in healthcare access.

The COVID-19 pandemic (Andrew, 2024) provided a stark example of how pseudoscience can fuel public health crises. False claims about unproven treatments—such as hydroxychloroquine, ivermectin, and various home remedies—led to widespread confusion and, in some cases, dangerous behaviors. Misinformation about the virus's origins, prevention measures, and vaccine safety exacerbated public fear and skepticism, ultimately hindering global efforts to control the spread of the disease.

The consequences of pseudoscience in health underscore the urgent need to promote scientific literacy (Romanova et al., 2024) and to ensure that public health policies and medical practices are grounded in robust, peer-reviewed evidence (Caron et al., 2024). Strengthening real science communication, encouraging critical thinking, and fostering trust in healthcare professionals are essential steps in mitigating the influence of pseudoscience and protecting both individual and public health.

CHAPTER 6

The Influence of Bias on Funding Mechanisms in Real Science.

The influence of bias on funding mechanisms in real science can significantly shape research outcomes and public perception. Bias, whether implicit or explicit, can affect the design, development, implementation, interpretation, and reporting of scientific studies, leading to skewed and erroneous results that may not accurately reflect reality. Funding mechanisms, particularly when tied to the science-industrial complex or specific political and economic agendas, can also create pressure to produce auspicious outcomes, potentially compromising the integrity of the research. This can lead to the selective publication of positive results, suppression of negative findings, or the prioritization of research areas that align with the funder's goals rather than broader societal needs. Ensuring transparency, robust subject matter expert peer review, and the independence of scientific inquiry is crucial to minimizing these influences and maintaining the credibility of real science (Roje et al., 2023).

6.1 How Cognitive Biases Affect Scientific Research.

6.1.1 Introduction.

Cognitive biases are systematic patterns of deviation from rationality that can unconsciously influence the way researchers perceive, interpret, and analyze data (Haselton et al., 2015). In scientific research, these biases can subtly affect decisions at various stages, from hypothesis formation to data interpretation, potentially leading to flawed conclusions or misguided research directions. Despite the rigorous methods designed to minimize errors, scientists, like all humans, are not immune to cognitive biases. Recognizing and mitigating these biases is essential to preserving the integrity and reliability of scientific findings (Macrina, 2014).

6.1.2 Types of Cognitive Biases in Scientific Research.

One common cognitive bias in scientific research is *confirmation bias*, where researchers may unintentionally favor information or data that supports their existing beliefs or hypotheses while overlooking or dismissing evidence that contradicts them. This can lead to selective data collection or interpretation, skewing results in a way that affirms the original hypothesis rather than objectively evaluating the evidence (Souchet et al., 2023).

Hindsight bias is another issue, where scientific researchers, after knowing the outcome of a study, may overestimate their ability to have predicted it, leading to an overconfidence in the correctness of their methods and findings (Kneer et al., 2023).

Publication bias occurs when studies with positive results are more likely to be published than those with null or negative results, which can distort the scientific literature by creating a false impression of consensus or efficacy (Kepes et al., 2023). This is often influenced by both cognitive biases of scientific researchers and the preferences of journals for "exciting" results.

Lastly, *anchoring bias* can also affect scientific research, where initial information or findings unduly influence subsequent interpretations or decisions, even when new evidence suggests a different conclusion (Rafiee-Rad et al., 2024).

6.1.3 Mitigating Cognitive Biases in Scientific Research.

To counteract cognitive biases that can distort scientific findings, it is crucial to implement research practices designed to minimize subjective influence and promote objectivity. One effective approach is the use of blind or double-blind study designs (Higgins et al., 2024), where information is concealed from both participants and researchers to reduce the impact of personal expectations or biases on the results. This method ensures that neither the participants' nor the researchers' prior beliefs influence the outcomes of the study.

Another important strategy is pre-registering studies (Petersen et al., 2024), a process in which researchers publicly declare their hypotheses, research questions, and methods before beginning the study. This practice helps to prevent selective reporting—where researchers might only publish data that supports their initial hypothesis—by holding them accountable to the full range of results, including those that may not align with their expectations.

Furthermore, encouraging the publication of all research results, regardless of whether they are positive, negative, or inconclusive, is vital for fostering scientific integrity. This approach reduces publication bias, which can distort the body of scientific knowledge by only showcasing studies that yield significant or expected findings. Cultivating a culture of transparency, truthfulness, and accountability in research—along with a commitment to replication studies—ensures that results are scrutinized and verified by others in the field. Replication studies are essential for confirming the reliability and validity of findings across different settings and populations, and they help prevent the persistence of erroneous conclusions in the scientific literature (Hardwicke et al., 2023).

By acknowledging the existence of cognitive biases and taking deliberate steps to mitigate their effects, the scientific community can enhance the credibility, rigor, and robustness of its inquiry. This not only strengthens the internal reliability of scientific knowledge but also ensures that the information presented to the public is accurate, transparent, and trustworthy, ultimately reinforcing public confidence in science.

6.2 The Impact of Funding Mechanisms on Scientific Research Outcomes.

6.2.1 Introduction.

Funding mechanisms for scientific research can significantly influence the direction, focus, and outcomes of studies (Munari et al., 2024; Van Thiel et al., 2024). While funding is essential for advancing knowledge and innovation, the motivations and interests of the funders can sometimes create conflicts of interest that affect the objectivity and integrity of scientific research (Periyasamy et al., 2024). Whether the funding comes from government agencies, private corporations, or nonprofit organizations, the expectations and priorities of these entities can shape the research agenda, potentially leading to biased outcomes that align with the funder's goals rather than purely scientific inquiry (Bendiscioli et al., 2023).

6.2.2 Scientific-Industrial Complex Influence on Scientific Research.

The scientific-industrial complex refers to the increasingly intertwined relationship between scientific research and industrial or corporate interests. This relationship can significantly influence the direction, priorities, and outcomes of scientific research. On one hand, the financial support from industries can provide essential resources for advancing research, particularly in fields like pharmaceuticals, biotechnology, and environmental science. These industries often fund large-scale projects and cutting-edge technologies that might be beyond the reach of public funding. However, this relationship can also lead to conflicts of interest, where scientific research is directed more by profit motives than by the pursuit of knowledge or the public good (Bowen et al., 2019; Cozzoli, 2024).

The influence of the scientific-industrial complex can slant research agendas towards areas that promise the most financial return, sometimes at the expense of more fundamental or exploratory real science that may not have immediate commercial applications. Moreover, there is a risk that the results of scientific research could be biased, with findings that are favorable to industry interests being more likely to be published or emphasized, while unfavorable results might be suppressed or downplayed. This dynamic can erode public trust in real science, as it blurs the line between objective inquiry and profit-driven outcomes (Briant, 2024). Addressing these challenges requires a careful balance of collaboration between industry and academia (Rossoni et al., 2024), transparency in funding sources (Adeusi et al., 2024), and a strong commitment to maintaining the integrity and independence of scientific research (Lovestam et al., 2024).

When scientific research is funded by private corporations, especially those with a financial stake in the outcomes, there is a risk that the studies may be designed or interpreted in ways that favor the sponsor's products or interests (Borsa et al., 2023). For example, pharmaceutical companies funding drug trials may have a vested interest in producing positive results, which can lead to selective reporting, manipulation of data, or even pressure on researchers to present findings in a favorable light (Kennedy, 2023). This can undermine the credibility of the research and lead to public skepticism, particularly when negative results are suppressed or downplayed. The influence of corporate funding can also skew the research landscape (Gomes et al., 2024), as studies with potential commercial applications may be prioritized over more fundamental or exploratory scientific research that lacks immediate profitability.

6.2.3 Government and Nonprofit Funding Mechanisms for Scientific Research.

While government and nonprofit organizations generally aim to support research for the public good, these funding sources can also influence research outcomes (Park et al., 2024). Political and ideological agendas can shape the allocation of government funds, directing research toward areas that align with current policy priorities while neglecting others (Hinterleitner et al., 2024). Nonprofit organizations, particularly those with specific missions or advocacy goals, may fund research that supports their causes, which can lead to biased interpretations or selective emphasis on certain findings (Ramos-Veilba et al., 2024). Additionally, competition for limited funding of scientific research can create pressure on researchers to produce results that are more likely to secure future grants, potentially compromising the objectivity of the research process itself (Mocanu et al., 2024).

6.2.4 Mitigating the Impact of Funding Bias in Scientific Research.

To minimize the influence of funding mechanisms on research outcomes, fostering transparency and upholding integrity in the scientific process are crucial. One essential step is the disclosure of funding sources and potential conflicts of interest, which enables greater scrutiny of the research process and helps maintain public trust in the validity and impartiality of the findings (Graham et al., 2024). When funding sources are transparent, it becomes easier for the scientific community and the public to evaluate whether the results may have been influenced by external financial interests, thereby safeguarding the integrity of the research.

In addition to financial transparency, subject matter expert peer review plays a pivotal role in ensuring the rigor and validity of scientific work. Peer review by independent experts provides an external check on the research design, methodology, and interpretations of data, helping to identify any flaws or biases that may have influenced the results. Replication studies are also vital, as they test the reliability of findings in different settings or by different research teams, reinforcing the robustness of scientific conclusions (Chou et al., 2024). Independent oversight, whether through institutional review boards, ethical committees, or external advisory panels, further ensures that research is conducted with the highest standards of accountability and integrity, free from undue influence by funding bodies.

Moreover, promoting open real science practices, such as making data and methodologies publicly accessible, enhances the transparency of the research process. This openness allows other scientists to verify results and fosters collaboration, leading to more rigorous and trustworthy outcomes. Encouraging diverse funding sources, including public, academic, and non-profit sectors, reduces the risk of bias associated with single-source funding and ensures that scientific research is not disproportionately influenced by the priorities or agendas of private companies or other interested parties.

By embedding these practices into the research process, the scientific community can mitigate the risk of bias and ensure that scientific outcomes are determined by empirical evidence rather than external financial or political pressures. This commitment to transparency and integrity not only strengthens the reliability of research findings but also reinforces the public's confidence in real science as an impartial and evidence-driven pursuit of knowledge.

6.3 Conflicts of Interest: Industry-Sponsored Studies vs. Independent Scientific Research.

6.3.1 Introduction.

Conflicts of interest arise when a researcher's personal, financial, or professional interests have the potential to compromise the objectivity and impartiality of their work. In scientific research, this issue becomes particularly concerning when industry-sponsored studies are compared to independent research (Graham et al., 2024). Industry sponsorship, especially from companies with a direct financial stake in the outcomes, can introduce significant biases. For example, pharmaceutical companies funding clinical drug trials or food manufacturers supporting nutrition studies may have motivations that align with favorable outcomes for their products. These financial connections can create a conflict of interest, leading to both subtle and overt pressures that could influence various aspects of the research process, from study design and data collection to the interpretation and reporting of results.

The influence of such conflicts can manifest in several ways. First, researchers may consciously or unconsciously design studies that are more likely to yield results that benefit their sponsors. This could involve selecting specific methodologies, populations, or endpoints that favor the product being tested, or even downplaying negative findings that could harm the sponsor's financial interests. Additionally, conflicts of interest may affect how data is interpreted or presented, potentially leading to biased conclusions that misrepresent the true effects of a drug, treatment, or product. For example, when results that are unfavorable to the sponsor's product are omitted or underreported, it skews the overall picture, giving an inaccurate portrayal of the efficacy or safety of the product.

Moreover, conflicts of interest can also affect the peer review process, as industry-sponsored studies may receive preferential treatment or scrutiny, depending on the researcher's relationship with the sponsor. This can lead to biased publishing decisions or a lack of rigorous evaluation, further undermining the credibility of the scientific findings. The public's trust in scientific research may erode if these biases are not adequately addressed, especially when industry-sponsored studies contradict independent research or when there is a lack of transparency about financial ties.

To mitigate the risks posed by conflicts of interest, it is crucial for researchers to disclose any financial relationships or personal interests that could influence their work. Journals, institutions, and funding agencies must enforce strict guidelines for transparency and accountability, ensuring that studies are conducted and reported with integrity. Independent research, free from external financial pressures, remains essential for maintaining the credibility of scientific inquiry and safeguarding the objectivity of research outcomes. This helps to ensure that findings are based on unbiased data and that the results serve the best interests of public health and well-being, rather than corporate agendas.

6.3.2 Industry-Sponsored Scientific Research Studies.

Industry-sponsored scientific research studies are frequently criticized for the potential biases they introduce, given that companies funding these studies have a vested interest in achieving favorable outcomes that align with the promotion of their products or services (Dubin et al., 2024). This financial stake can lead to various practices that undermine the objectivity and credibility of the research. One common issue is selective reporting (Huang, 2024), where only positive results are published, while negative or inconclusive findings are either downplayed or omitted entirely. This selective dissemination of data creates a skewed representation of the product's effectiveness or safety, which can mislead the public and healthcare professionals.

Another concern is the framing of conclusions (Graham et al., 2024), where researchers, knowingly or unknowingly, present findings in a way that minimizes risks or downplays negative effects. For example, potential side effects or harmful outcomes may be underemphasized, leading to an overly optimistic portrayal of a product. Such selective interpretation of data can severely affect decision-making in both clinical and regulatory settings, where the understanding of a product's risks and benefits is critical.

Industry sponsors may also exert influence over the design of the study itself (Traeger et al., 2024), including the choice of control groups, dosing regimens, or statistical methods. These design decisions, when tailored to favor the sponsor's interests, can skew the results in favor of the desired outcome. For instance, an industry-sponsored study may use a control group that is not representative of the general population, or it may employ dosing schedules that optimize the likelihood of positive results. Similarly, the statistical methods employed may be chosen to highlight favorable results while obscuring any significant negative findings.

The pressure to deliver outcomes that align with the sponsor's objectives can also impact the behavior of researchers, both consciously and unconsciously, leading to conflicts of interest that compromise the integrity of the research (Crawford, 2024). Researchers may feel incentivized to produce results that support the sponsor's product, especially when funding or professional advancement is at stake. This creates an environment where the objectivity of the research is compromised in favor of achieving a desired outcome.

As a result, industry-sponsored studies are often viewed with skepticism, particularly when their findings contradict those from independent, unbiased research. This skepticism is further exacerbated when industry-sponsored research is presented without adequate transparency regarding potential conflicts of interest or financial ties. To maintain trust in scientific inquiry and protect the integrity of the research process, it is essential that such studies undergo rigorous scrutiny, and that transparency regarding funding sources and conflicts of interest is prioritized. By ensuring that studies are designed, conducted, and reported with the highest standards of integrity, the scientific community can mitigate the risks associated with industry sponsorship and maintain the credibility of research outcomes.

6.3.3 Independent Scientific Research.

Independent scientific research, typically funded by government agencies, nonprofit organizations, or academic institutions, is often viewed as more objective and less influenced by commercial interests compared to industry-sponsored studies. This perception arises from the fact that researchers conducting independent studies are generally motivated by scientific curiosity and the pursuit of knowledge rather than financial gain. This intrinsic motivation can reduce the likelihood of biases that might arise when there are financial stakes tied to the outcomes, as is the case with industry-sponsored research. The absence of direct profit-driven interests allows independent research to prioritize the integrity of the scientific process, fostering greater confidence in the results and their applicability to the broader scientific community.

However, it is important to recognize that even independent research is not entirely free from conflicts of interest. Researchers may still face pressures that could influence their objectivity. These pressures can include the desire for career advancement, the ambition to publish in prestigious journals, or the need to secure future research funding. In some cases, researchers may feel compelled to produce results that align with the interests of funding bodies or institutions to ensure continued support for their work. This is particularly relevant when funding is competitive or tied to specific research outcomes, such as in certain areas of applied science or medical research.

Despite these potential conflicts, independent research tends to be more transparent and open to scrutiny than industry-sponsored studies. One of the key benefits of independent research is the relative freedom researchers have in sharing data and methodologies, which allows other scientists to replicate, validate, or challenge the findings. This transparency helps mitigate biases and increases the reliability of the results, as the broader scientific community can independently verify the research process and outcomes (Wu et al., 2024; Forscher et al., 2023). Independent research is also subject to more rigorous peer review, where experts in the field critically evaluate the study's design, methodology, and conclusions before publication, further ensuring the credibility and quality of the work.

Moreover, academic and government-funded research often operates within a framework that prioritizes ethical standards and accountability, with clear guidelines for data sharing, conflict of interest disclosures, and publication practices. These safeguards help reduce the risk of research being skewed by outside influences and promote a culture of integrity and collaboration.

Ultimately, while independent research may still face some of the challenges related to bias and conflicts of interest, it remains an essential pillar of scientific progress. The greater levels of transparency, subject matter expert peer review, and freedom from commercial pressures help ensure that independent research contributes to a more accurate and reliable understanding of the natural world.

6.3.4 Balancing Industry and Independent Scientific Research.

To effectively address conflicts of interest, it is crucial to strike a balance between industry-sponsored and independent scientific research (Sessler et al., 2023). Industry funding plays an essential role in advancing certain areas of scientific inquiry, particularly those requiring substantial financial investment, such as drug development, advanced technologies, and large-scale clinical trials. However, the involvement of industry introduces potential biases that can affect the objectivity and integrity of the research. To mitigate these risks, it is essential that industry-sponsored studies adhere to strict guidelines and transparency measures to ensure that the research process remains impartial and trustworthy.

One of the key components in maintaining this balance is the full disclosure of funding sources and any potential conflicts of interest. Transparency is critical in allowing the scientific community and the public to assess the potential influences that could shape the research outcomes. By openly identifying financial backers and potential conflicts, researchers and institutions can increase accountability and minimize the risk of bias. Furthermore, independent oversight and peer review should be integral to the research process, providing an additional layer of scrutiny to ensure the validity and reliability of the findings. Peer review by subject matter experts in the field helps to identify flaws in study design, methodology, and interpretation, and ensures that the results meet the highest standards of scientific rigor.

In addition to transparency and subject matter expert peer review, fostering collaboration between industry-sponsored researchers and independent researchers can further help to mitigate biases. Such partnerships can combine the strengths of both sectors, where the resources and expertise from industry can complement the critical thinking and objectivity inherent in academic and independent research. For instance, independent researchers can bring a fresh, unbiased perspective to industry-sponsored studies, offering constructive critiques or alternative interpretations that may not align with commercial interests but could still lead to valuable insights. These collaborative efforts can also encourage the sharing of data, methodologies, and results across different research communities, promoting openness and accelerating the advancement of scientific knowledge.

Moreover, promoting diversity in funding sources and encouraging research initiatives that are independent of corporate interests can help prevent any single entity from having undue influence over the direction of scientific inquiry. It is also vital to establish regulatory frameworks and ethical guidelines that govern industry-sponsored research, ensuring that studies are conducted with transparency and integrity and that any potential conflicts are actively managed.

Ultimately, maintaining a balance between industry-sponsored and independent research is key to advancing real science while protecting its objectivity. Through transparency, independent oversight, and collaboration, the scientific community can help ensure that research remains credible, reliable, and driven by empirical evidence rather than external commercial interests. This approach fosters a more robust and trustworthy scientific process, which is essential for achieving meaningful and ethical advancements in science and medicine.

CHAPTER 7

The Reproducibility Crisis.

The reproducibility crisis in real science refers to the growing concern that many scientific studies cannot be replicated or reproduced, undermining the reliability and validity of research findings (Fisar et al., 2024). This crisis has been highlighted by numerous high-profile failures to replicate results across various fields, from psychology to medicine. Factors contributing to the reproducibility crisis include small sample sizes, selective reporting of positive results, inadequate methodological transparency, and statistical errors (Klonsky, 2024). The lack of reproducibility challenges the strength of scientific knowledge and emphasizes the need for improved research practices, such as rigorous subject matter expert peer review, open data sharing, and replication studies, to ensure that scientific claims are valid and reliable (Stoudt et al., 2024).

7.1 Understanding the Reproducibility Crisis in Scientific Research.

7.1.1 Introduction.

The reproducibility crisis in scientific research is a critical issue within the scientific community, marked by challenges in replicating the results of published studies (Antunes et al., 2024). This problem spans multiple scientific disciplines and raises concerns about the reliability of research findings. The crisis arose when independent researchers attempted to replicate experiments using the original methods and data, only to find that many results could not be reproduced. This failure to achieve reproducibility threatens the core principles of scientific inquiry, which depend on the ability to verify and validate findings through repeated experimentation and observation.

7.1.2 Contributing Factors to the Reproducibility Crisis.

Several interrelated factors contribute to the reproducibility crisis, posing significant challenges to the integrity of scientific research (Hicks, 2023). One of the most prominent issues is the widespread use of small sample sizes, which can produce statistically weak or non-generalizable results. Studies with insufficient sample sizes are more susceptible to random variation, increasing the likelihood of false-positive findings and reducing the reliability of conclusions drawn from the data.

Another major factor is selective reporting and publication bias, where studies with positive or novel findings are more likely to be published, while those with null or negative results are often disregarded (Kepes et al., 2023). This distortion of the scientific literature creates an illusion of stronger evidence than what actually exists, as unsuccessful replication attempts and contradictory findings frequently go unpublished.

Inadequate methodological transparency further exacerbates the crisis, making it difficult for other researchers to replicate and validate published findings (Hardwicke et al., 2023; Graham et al., 2024). When studies lack clear and detailed documentation of their methods, data collection procedures, and analytical techniques, independent researchers struggle to reproduce the original experiments accurately. This lack of transparency may stem from unintentional omissions, poor research practices, or, in some cases, deliberate obfuscation.

Additionally, the intense pressure to publish in high-impact journals incentivizes researchers to prioritize novel, striking, or sensational findings over methodologically rigorous and replicable science (Mocanu et al., 2024). This "publish or perish" culture encourages practices such as p-hacking (manipulating data to achieve statistical significance), HARKing (hypothesizing after results are known), and the selective presentation of data, all of which contribute to irreproducible research. Addressing these systemic issues requires cultural and structural changes in how scientific research is conducted, reviewed, and disseminated.

7.1.3 Implications and Solutions to the Reproducibility Crisis.

The implications of the reproducibility crisis are far-reaching, impacting not only the advancement of scientific knowledge but also public trust in research and its applications (Schmidt et al., 2024; Cole et al., 2024). When findings cannot be reliably replicated, confidence in the validity of scientific claims diminishes, potentially leading to misguided policies, wasted resources, and the adoption of ineffective—or even harmful—practices based on flawed data. In fields such as medicine, psychology, and environmental science, irreproducible results can have severe consequences, influencing clinical practice guidelines, public health interventions, and healthy public policy decisions that affect millions of lives.

Addressing this crisis requires a multifaceted approach that strengthens the foundations of research integrity, transparency, and accountability (Graham et al., 2024). One key strategy is improving research practices through more rigorous study design, appropriate statistical analyses, and greater methodological transparency. Encouraging preregistration of studies—where researchers publicly document their hypotheses, methodologies, and analytical plans before data collection—helps reduce bias and increases the credibility of findings (Petersen et al., 2024).

Additionally, promoting open science initiatives, such as data and code sharing, fosters greater reproducibility by enabling independent researchers to scrutinize and verify results. Journals and funding agencies can play a crucial role by incentivizing transparency and rewarding replication efforts rather than solely prioritizing novel discoveries.

Creating a research culture that values replication studies, rigorous subject matter expert peer review, and methodological soundness over sensationalism is essential for restoring both the publics and the scientific community's trust in real science (Chou et al., 2024). This shift requires academic institutions, funding bodies, and publishers to recognize and support replication efforts, ensuring that real science remains self-correcting and continues to contribute meaningfully to human progress.

7.2 Factors Contributing to Irreproducible Results in Scientific Research.

7.2.1 Introduction.

Irreproducible results—scientific findings that cannot be reliably replicated by independent researchers—pose a major challenge to the credibility and progress of science (Jarvis, 2024). One of the primary contributors to this issue is the use of small sample sizes, which can undermine the statistical power and generalizability of research findings (Cao et al., 2024). When studies rely on limited sample sizes, they are more susceptible to random fluctuations and sampling errors, making it difficult to distinguish true effects from chance occurrences.

Small sample sizes also contribute to inflated effect sizes, as extreme values are more likely to emerge in smaller datasets. This can create a false perception of strong associations, leading researchers and policymakers to overestimate the significance of findings. Furthermore, studies with inadequate sample sizes have a higher probability of yielding false positives—results that appear statistically significant but are not reproducible in subsequent experiments with larger or more diverse populations.

The consequences of these statistical weaknesses extend beyond academic discourse. In fields such as medicine, psychology, and social sciences, policy decisions and clinical practice guidelines based on small, irreproducible studies can mislead practitioners, waste resources, and even cause harm to the public-at-large. Addressing this issue requires a shift toward more rigorous study designs, including power analyses to ensure adequate sample sizes, replication efforts to confirm findings, and transparency in reporting data collection and analysis methods. Encouraging larger, multi-center collaborations and promoting open science initiatives can further help mitigate the risks associated with small sample sizes, ultimately strengthening the reliability of scientific research.

7.2.2 Methodological Issues Contributing to Irreproducible Results in Scientific Research.

Methodological shortcomings are a key factor contributing to irreproducibility, as they can compromise the accuracy and reliability of scientific findings (Sikorski, 2024). One major issue is the inadequate or poorly documented description of experimental procedures, which makes it difficult for other researchers to replicate studies with precision (Tang, 2024). When protocols lack sufficient detail, small but significant variations—such as differences in sample preparation, reagent concentrations, or data processing techniques—can lead to discrepancies in results. Even minor inconsistencies in how experiments are conducted or how data is collected can undermine replication efforts, causing confusion and misinterpretation.

Moreover, methodological flexibility—the ability of researchers to make multiple subjective decisions throughout the research and analysis process—can introduce bias and distort findings. This flexibility increases the risk of *p-hacking*, a problematic practice in which researchers manipulate data, selectively report findings, or adjust statistical analyses to achieve statistically significant results (Goel et al., 2024). For example, researchers may test multiple hypotheses, alter exclusion criteria, or selectively report only favorable results without acknowledging failed analyses. These practices, whether intentional or unintentional, inflate the likelihood of false positives and contribute to the growing body of irreproducible research.

Addressing these methodological challenges requires a commitment to greater transparency, accountability, and standardization in scientific research. Previously discussed, encouraging the use of pre-registered study designs, where researchers define their hypotheses, methods, and analysis plans before data collection, can help mitigate the risks of selective reporting and analytical bias. Additionally, promoting open science initiatives—such as sharing raw data, code, and detailed protocols—enhances the ability of independent researchers to verify and replicate findings. Greater adherence to standardized reporting guidelines and stronger subject matter expert peer review processes can further help ensure that scientific studies are conducted rigorously and contribute meaningfully to the advancement of knowledge.

7.2.3 Publication Bias and Pressure Contributing to Irreproducible Results in Scientific Research.

Publication bias and the pressure to produce novel, statistically significant results are major drivers of irreproducibility in scientific research (Keyes et al., 2023). The academic reward system often prioritizes groundbreaking discoveries over rigorous, incremental contributions, leading researchers to selectively report results that appear novel or confirm hypotheses while downplaying or omitting negative or null findings. This selective reporting distorts the scientific literature, creating a publication landscape that overrepresents positive results and underrepresents studies that fail to find significant effects.

One of the key consequences of publication bias is the false inflation of the perceived validity and effectiveness of certain findings. When negative or null results are systematically excluded from publication, the scientific community may unknowingly rely on incomplete or misleading evidence. Previously discussed, this bias can have serious implications, particularly in fields such as medicine, psychology, and social sciences, where clinical guidelines, policy decisions, and public health strategies are often based on published research. If only successful interventions or treatments are widely reported, ineffective or even harmful practices may be adopted, leading to wasted limited resources and potential risks to patient safety.

Addressing publication bias requires structural changes in how research is evaluated and disseminated. Encouraging journals to accept studies based on methodological rigor rather than the significance of results, increasing the publication of replication studies, and promoting platforms for sharing null results are essential steps toward a more balanced and accurate scientific record. Referred to earlier, initiatives such as registered reports—where journals commit to publishing studies based on their methodology rather than their outcomes—can help shift the focus from sensational findings to robust, reproducible science. Expanding open-access data repositories and fostering a culture that values transparency over novelty will further strengthen the credibility of scientific research.

7.2.4 Statistical Issues Contributing to Irreproducible Results in Scientific Research.

Statistical issues are a major contributor to irreproducible results, often compromising the reliability and validity of scientific findings (Yeo-The et al., 2024). One of the most common problems is inadequate statistical power, which arises when studies use small sample sizes or insufficient controls. Low-powered studies are more susceptible to random fluctuations, increasing the likelihood of both false positives (Type I errors) and false negatives (Type II errors). As a result, findings from underpowered studies may not hold up when tested with larger or more diverse populations, reducing their generalizability and scientific value.

Another critical issue is the misuse or misunderstanding of statistical methods, which can distort results and lead to incorrect conclusions (Goel et al., 2024). Many researchers rely heavily on p-values as the primary measure of statistical significance, often misinterpreting them as definitive proof of an effect. However, p-values are highly sensitive to sample size and do not indicate the magnitude or practical importance of an effect. Over-reliance on arbitrary thresholds (such as $p < 0.05$) encourages questionable research practices like p-hacking, where researchers manipulate analyses to achieve statistical significance, even if the results are not meaningful.

Failure to account for multiple comparisons further exacerbates the problem. When researchers conduct multiple statistical tests without adjusting for multiple hypothesis testing, the probability of false positives increases significantly. This is particularly problematic in large datasets and exploratory research, where conducting many comparisons without proper statistical corrections can yield misleading results. Similarly, improper use of statistical modeling, such as overfitting (creating models that fit the specific dataset but fail to generalize) or inappropriate handling of confounding variables, can further undermine reproducibility.

Addressing these statistical challenges requires greater emphasis on rigorous study design and appropriate analytical techniques. Researchers should conduct power analyses before data collection to ensure adequate sample sizes and use effect sizes and confidence intervals alongside p-values to provide a more comprehensive understanding of findings. Adopting Bayesian statistical approaches, pre-registering analysis plans, and promoting statistical education within the scientific community can help reduce errors and improve the reproducibility of research. Journals and reviewers also play a crucial role in encouraging proper statistical reporting and discouraging common pitfalls, ultimately strengthening the credibility of scientific inquiry.

7.3 Solutions for Mitigating Irreproducibility: Transparency, Open Data, and Better Statistical Practices.

The reproducibility crisis in scientific research is a significant problem, undermining both the scientific community's and the public-at-large's trust in scientific findings and slowing technological progress. A robust solution to this crisis lies in the integration of transparency, open data, and better statistical practices.

Transparency involves making the entire research process—from hypothesis formulation to data collection and analysis—open and accessible to other researchers (Hardwicke et al., 2024). With open access to methodologies, protocols, and even raw data, researchers enable others to understand, replicate, or build upon their work. This transparency not only enables the verification of results but also promotes accountability while fostering active collaboration, coordination, and cooperation across disciplines and sectors. By encouraging interdisciplinary and multisectoral engagement, it strengthens the scientific process, leading to more robust and reliable discoveries.

Open data is a central part of this transparency (Chakravorti et al., 2024). By making data publicly available, researchers empower the scientific community to reanalyze results, apply different statistical methods, or explore new hypotheses using the same dataset. This practice enhances the reproducibility of findings, as it allows for the independent verification of results. Moreover, open data can lead to new insights and innovations that were not originally anticipated, as other researchers may approach the data from different angles or with novel tools.

Better statistical practices are also essential in addressing the reproducibility crisis (Antunes et al., 2024; Fisar et al., 2024; Petersen et al., 2024). Previously discussed, the misuse of statistical methods, such as p-hacking or failing to account for multiple comparisons, has contributed to the problem. Implementing more rigorous statistical standards, such as pre-registering studies, using appropriate sample sizes, and emphasizing effect sizes over p-values, can significantly improve the reliability of research findings. Additionally, promoting the use of advanced statistical techniques and encouraging the publication of negative or null results can help to create a more accurate and balanced scientific record.

By combining transparency, open data, and better statistical practices, the scientific community can move towards more reproducible and trustworthy research, ultimately accelerating scientific progress.

CHAPTER 8

The Global Scientific-Industrial Complex: Impact on the Scientific Enterprise.

The global scientific-industrial complex refers to the collaboration between scientific research institutions and commercial industrial enterprises world-wide, where commercial interests increasingly influence the direction and funding of scientific research (Kirillova, 2020). This relationship can drive innovation and accelerate the development of new advanced medical and information technologies, as commercial industry support provides necessary resource allocation, funding mechanisms, and practical applications for scientific discoveries. However, it also raises concerns about potential conflicts of interest, where research priorities may shift towards profitable ventures rather than addressing fundamental scientific questions or public needs (Thurik et al, 2024)

The impact on global scientific enterprise includes both the potential for groundbreaking advancements and the risk of compromising scientific integrity, as mis-aligned financial incentives may influence research outcomes and access to data (Novelli et al., 2024). Balancing these interests is crucial to ensuring that scientific progress remains transparent, accountable, ethical, and aligned with broader societal goals (Sessler et al., 2023).

8.1 The Institutional Aspects of Real Science.

The institutional aspects of real science refer to the structures, organizations, and norms that support and regulate global scientific activity (Erduran, 2023). These institutions play a crucial role in shaping the practice of real science, providing the resources, frameworks, and oversight necessary for transparent, valid, and reliable scientific research and development. They include universities, research institutes, funding agencies, professional societies, and governmental bodies, all of which contribute to the advancement of scientific knowledge and the integration of real science into society.

Universities and research institutes are fundamental to the institutional framework of real science (Clark, 2023). These institutions provide the infrastructure, funding, and collaborative environment necessary for conducting scientific research. They also play a key role in education and training, preparing the next generation of scientists through rigorous academic programs and hands-on structured research experiences. By fostering a culture of inquiry and innovation, universities and research institutes help maintain the vitality and progress of scientific disciplines.

Funding agencies, both public and private, are vital for supporting scientific research (Munari et al., 2024). These organizations allocate financial resources to projects and researchers, enabling the pursuit of scientific questions that might not have immediate commercial applications but are essential for long-term advancement. Government agencies such as the National Science Foundation (NSF) in the US or the European Research Council (ERC) in the EU provide substantial funding for basic and applied research, reflecting societal priorities and national interests. Private foundations and corporations also contribute significantly, often targeting specific fields or problems. The availability and distribution of funding can significantly influence the direction, scope, and scale of scientific research (Ramos-Vielba et al., 2024).

Professional societies and associations, such as the American Association for the Advancement of Science (AAAS) or the Royal Society, also play an important role in the global institutional landscape of real science (Sutter, 2024). These organizations provide platforms for the global dissemination of research findings through conferences, journals, and publications. They also establish professional standards, codes of ethics, and best research practices, ensuring the integrity and quality of scientific work. By fostering professional networks and communities, these societies facilitate collaboration, mentorship, and the exchange of ideas among scientists.

Governmental bodies and regulatory agencies around-the-world oversee various aspects of scientific research and its applications, ensuring that they meet ethical, safety, and environmental standards (Choudhary et al., 2024). National agencies, such as the Food and Drug Administration (FDA) in the US, regulate the development and approval of new or improved medical treatments and advanced medical technologies, ensuring they are safe and effective for public use (Watson et al., 2023). International and national environmental protection agencies (EPAs) oversee the impact of scientific and technological activities on the environment, promoting sustainable practices (Glicksman et al., 2023). These global regulatory frameworks are essential for maintaining public trust and ensuring that scientific advancements benefit society as a whole.

The institutional aspects of real science also encompass the cultural and societal norms that guide scientific practice (Gauchat, 2023). Norms of transparency, subject matter expert peer review, and open communication are foundational to the credibility and progress of real science. The peer review process plays a crucial role in ensuring that scientific research undergoes rigorous scrutiny and validation by subject matter experts, both within the field and, at times, from complementary sectors. This critical evaluation helps maintain the quality, reliability, and credibility of published findings. Open communication and data sharing promote collaboration and enable other researchers to build on existing work, accelerating scientific progress.

In summary, the institutional aspects of real science worldwide, encompasses the diverse structures, organizations, and norms that support and regulate scientific activity. Universities and research institutes, funding agencies, professional societies, and governmental bodies all play critical roles in advancing scientific knowledge and integrating it into society. These institutions provide the resources, frameworks, and oversight necessary for conducting high-quality scientific research, ensuring that scientific advancements are ethical, equitable, and safe to society.

8.2 Characteristics of a Strong and Sustainable Global Scientific Enterprise.

A strong and sustainable global scientific enterprise is built upon several foundational elements that ensure its robustness, resilience, and continuous innovation (Nightingale et al., 2024). These elements include a commitment to scientific integrity, supportive infrastructures, effective collaboration, stable and diversified funding, public engagement, emphasis on education and training, adaptability, and a culture of innovation. Together, these factors create an environment conducive to high-quality research and long-term societal benefits.

At the heart of a strong global scientific enterprise is a steadfast commitment to scientific integrity and ethical standards (Pols et al., 2024). This involves rigorous adherence to principles of honesty, transparency, and accountability in all aspects of research. Ethical guidelines ensure that experiments are conducted responsibly, considering the welfare of human and animal subjects and the environment. Upholding high ethical standards not only enhances the credibility of research findings but also builds public trust in science (Kishan et al., 2024), which is essential for the continued support and relevance of scientific endeavors.

A strong infrastructure is critical for a thriving global scientific enterprise (Radberg et al., 2024). This includes state-of-the-art laboratories, advanced scientific technologies, and access to high-quality research facilities. Such infrastructure enables researchers world-wide to conduct experiments with precision and accuracy, facilitating and sharing groundbreaking discoveries. Institutional support in the form of administrative services, grant management, and regulatory compliance further streamlines research operations, allowing scientists to focus on their investigative work.

Global collaboration and communication across disciplines, institutions, and borders is vital for addressing complex scientific questions and societal challenges (Bowser et al., 2024). Interdisciplinary and multisectoral research fosters innovation by integrating diverse perspectives and expertise, leading to more comprehensive solutions. Collaborative networks facilitate the sharing of resources, data, information, and knowledge, enhancing the overall productivity and impact of scientific research. By promoting teamwork and cross-pollination of ideas, scientific enterprises can tackle multifaceted problems more effectively.

Sustained financial support is essential for the continuity and growth of scientific research (Saldana, 2024). Adequate and stable funding allows researchers to pursue ambitious projects, invest in cutting-edge technologies, and attract top talent. Diversified funding sources, including international and national governmental agencies, private foundations, and industry partnerships, help mitigate the risks associated with economic fluctuations and policy changes. A strong funding base ensures that scientific enterprises can plan long-term research agendas and adapt to emerging challenges.

Engaging with the public and effectively communicating scientific findings world-wide are crucial for fostering a supportive environment for real science (Bayes et al., 2023). Public understanding and appreciation of real science can lead to increased support for research funding and informed decision-making. Scientists must actively communicate their work through various channels, including media, public lectures, and educational outreach programs. Transparency in communicating the goals, methods, and implications of research helps build trust and encourages public participation in the scientific process (e.g., Citizen Scientist, Community-based Participatory Research (CBPR)).

A continuous pipeline of well-trained scientific researchers and professionals is vital for a sustainable global scientific enterprise (Ladner et al., 2024). Investing in real science education at all levels, from primary school to postgraduate training, is essential for developing the next generation of scientists. Education programs should emphasize critical thinking, problem-solving, logical reasoning, and hands-on research experiences. Additionally, ongoing professional development opportunities help researchers stay current with advancements in their fields and develop new skills, ensuring the scientific workforce remains dynamic and capable.

The ability to adapt to changing circumstances and embrace innovation is crucial for the long-term sustainability of global scientific enterprises (Mosteanu et al., 2024). This includes staying abreast of new or improved technological advancements, emerging research methodologies, and shifting societal needs, preferences, and values. Encouraging a culture of creativity and innovation, where risk-taking and creative problem-solving are valued, helps global scientific enterprises remain dynamic and responsive. Institutions should also be flexible in their organizational structures, processes, and policies to accommodate new directions in research, education, and service.

In summary, a strong and sustainable global scientific enterprise is built on a foundation of integrity, ethical conduct, and robust infrastructure. It thrives through effective collaboration, stable funding, public engagement, and a commitment to education and training. Additionally, fostering a culture of adaptability and innovation is essential for continuous advancement. By nurturing these key elements, global scientific enterprises can not only endure but also excel, driving the discovery of new knowledge and addressing society's most complex challenges. This holistic, interdisciplinary, and multisectoral approach ensures that scientific research remains impactful, resilient, and sustainable, shaping progress for generations to come.

8.3 The Effects of the Global Scientific-Industrial Complex on the Global Enterprise of Real Science.

The global Scientific-Industrial Complex, characterized by the symbiotic relationship between scientific research and commercial industrial interests, has profound effects on the global enterprise of real science (Ghobakhloo et al., 2023). This relationship drives significant advancements by leveraging industry resources for scientific innovation, while also posing challenges related to the direction, ethics, and integrity of research. Understanding these effects is crucial for fostering a balanced and sustainable global scientific enterprise.

One of the most notable positive effects of the Scientific-Industrial Complex is its role in driving innovation and advancements (Mosteanu et al., 2024). Industry funding and resource allocation enable large-scale research projects, the development of cutting-edge technologies, and the acceleration of scientific discoveries (Saldana, 2024). This partnership often results in practical applications that benefit society world-wide, such as new medical treatments, advanced materials, and sustainable technologies. The collaborative efforts between industry and academia (Rossoni et al., 2024) create an environment where scientific ideas can be rapidly translated into real-world solutions, promoting technological progress and economic growth.

However, the Scientific-Industrial Complex can also influence research priorities in ways that may not always align incentives with broader scientific or societal goals (Krishna, 2024). Industrial funding often focuses on projects with clear commercial potential, which can lead to an emphasis on applied research over fundamental or exploratory real science. This shift in priorities may result in underfunding of basic research that, while not immediately profitable, is essential for long-term scientific breakthroughs and understanding. Ensuring a balance between industry-driven research and independent, curiosity-driven real science is crucial for maintaining a healthy and comprehensive global scientific enterprise.

The close ties between industry and real science raise important ethical considerations and potential conflicts of interest (Pols et al., 2024). Financial dependencies can sometimes compromise the objectivity and integrity of scientific research. For example, studies funded by industry may be biased towards favorable outcomes (Cantner et al., 2024), or there may be pressure to withhold negative results (Parker et al., 2024). These conflicts of interest can undermine public trust in scientific findings and erode the credibility of research institutions. Implementing stringent ethical guidelines and transparent disclosure practices is essential to mitigate these risks and uphold the integrity of scientific research.

On a more positive note, the Scientific-Industrial Complex often leads to enhanced infrastructure and research capabilities (Radberg et al., 2024). Industrial partnerships provide access to advanced medical and information technologies, sophisticated laboratories and research institutions, and specialized subject matter expertise that may not be available in all academic settings around-the-world. These resources enable scientists to conduct high-quality research and tackle complex problems more effectively. Collaborative projects can also foster skill development and knowledge transfer between industry and academia (Rossoni et al., 2024), enriching the scientific workforce and promoting interdisciplinary and multisectoral strategies and approaches.

The Scientific-Industrial Complex significantly impacts public perception of real science and influences real science policy (Liu, 2024). When industry and real science work together transparently, truthfully, and ethically, they can enhance public confidence in scientific research and its benefits. Conversely, instances of misconduct or perceived conflicts of interest can lead to skepticism and distrust. Policymakers often rely on scientific advice shaped by commercial industry interests, which can affect regulatory decisions, funding and resource allocations, and research agendas. Ensuring that real science policy is informed by unbiased and rigorous valid and reliable research is vital for addressing societal challenges and promoting public welfare (Cochrane et al, 2024).

Finally, the Scientific-Industrial Complex plays a key role in promoting economic growth and national competitiveness (Lu, 2024). Collaborative research initiatives contribute to the development of new industries, job creation, and the commercialization of innovative products. Countries that effectively integrate real science and industry are often leaders in advanced technology and innovation, enhancing their global economic standing. However, this competitive drive should not overshadow the need for responsible and sustainable scientific practices that prioritize long-term societal benefits over short-term economic gains.

In summary, the Scientific-Industrial Complex has a complex and sometimes complicated impact on the enterprise of real science. While it drives innovation, enhances research capabilities, and promotes economic growth, it also presents challenges related to research priorities, ethical considerations, and public trust. Balancing these effects through transparent practices, ethical guidelines, and a commitment to independent and fundamental research is essential for fostering a strong and sustainable scientific enterprise that serves the broader interests of society.

8.4 The Scientific World View Under the Enterprise of Real Science.

The scientific worldview is a fundamental aspect of the enterprise of real science, shaping how scientists understand and interpret the natural world (Matthews, 2009; Abd-El-Khalick, 2012). This worldview is characterized by a set of principles that guide scientific inquiry and distinguish it from other ways of knowing. Emphasizing empirically-derived evidence, logical reasoning, critical thinking, and skepticism, the scientific worldview underpins the methods and practices that drive scientific discovery and modern technological advancement.

Central to the scientific worldview is the reliance on empirical-ly-derived evidence, experimentation, and observation (Allchin, 2011). Scientists gather data through systematic observations, experiments, and measurements to construct a clear, comprehensible and truthful understanding of phenomena. This empirically-driven evidence-based approach ensures that scientific knowledge is grounded in observable reality and can be tested and verified by others. By prioritizing empir-ically-derived evidence, scientists can build a body of information and knowledge that is objective, valid, reliable, reproducible, and continu-ally open to revision based on new findings in real time.

The scientific worldview also emphasizes logical reasoning and critical thinking (Les, 2024; Desmond et al., 2024). Scientists use deductive and inductive reasoning to form hypotheses, design experiments, and draw conclusions. This logical framework allows for the formulation of testable predictions and the systematic evalu-ation of results. Critical thinking enables scientists to scrutinize their own assumptions, identify potential biases, and rigorously assess the validity of their findings. This disciplined scientific approach fosters a deeper understanding of complex and complicated issues and pro-motes the development of robust theories, frameworks, and models.

Another key aspect of the scientific worldview is the balance between skepticism and open-mindedness (Youvan, 2024). Scientists are trained to question assumptions, challenge established ideas, and critically evaluate new information. This skepticism helps to guard against errors, biases, and unsubstantiated claims. At the same time, scientists must remain open-minded, willing to consider alternative explanations and revise their views in light of new empirically-de-rived evidence. This dynamic interplay between skepticism and open-mindedness drives scientific progress and ensures that informa-tion and knowledge evolve in a self-correcting manner.

The scientific worldview is rooted in the belief that the laws of nature are universal and consistent (Shorina et al., 2024). This means that scientific principles apply uniformly across different contexts and can be used to explain phenomena in multisectoral disciplines.

For example, the laws of physics govern the behavior of objects on Earth as well as in distant galaxies (Chen, 2025). This universality allows scientists to make broad generalizations and develop theories that have wide-ranging applicability. It also facilitates the integration of knowledge across various scientific disciplines, promoting a holistic, multisectoral understanding of the natural world.

The scientific enterprise is fundamentally collaborative, depending on the collective efforts of researchers, institutions, and societies (Schubert, 2024). This worldview emphasizes the value of sharing information, knowledge, expert peer review, and building upon the work of others to ensure sustainability and growth. Collaborative networks allow scientists to combine resources, exchange ideas, and validate results through independent replication.

This communal aspect of real science enhances the robustness of scientific knowledge and accelerates the pace of discovery (Enquist et al., 2024). It also underscores the role of the scientific community in fostering a culture of inquiry, integrity, and continuous learning (Pols et al., 2024).

Finally, the scientific worldview encompasses a strong ethical dimension, emphasizing the responsibility of scientists to conduct their work with integrity and to consider the broader implications of their research (Pols et al., 2024). Ethical guidelines ensure that scientific practices are conducted humanely and responsibly, with respect for human and animal subjects' well-being, the environment, and the public-at-large. Scientists are also tasked with communicating their findings transparently and truthfully and engaging with the public to promote informed decision-making. This ethical commitment is crucial for maintaining public trust in real science and ensuring that scientific advancements contribute positively to society (Gunderson, 2024).

In summary, the scientific worldview is grounded in an empirically-driven evidence-based approach emphasizing the fulfillment of responsibilities through well-established principles such as critical thinking, logical reasoning, skepticism, open-mindedness, universality, collaboration, and ethical responsibility. These core values shape the methods and practices of scientific inquiry, fostering a dynamic, self-correcting process that continually expands our understanding of the natural world. By adhering to this worldview, scientists can strengthen and sustain a thriving scientific enterprise, driving creativity and innovation, tackling complex and complicated problems, and improving the well-being of all living things.

8.5 The Scientific Method of Inquiry Under the Enterprise of Real Science.

The scientific method of inquiry (Anderson et al., 2024) is an organized approach used by scientists to investigate natural phenomena, acquire knowledge, and develop theories about the world around us. This method is characterized by its thoroughness, fairness, and dependence on empirically-derived evidence, distinguishing it from other forms of inquiry. While there is no single fixed scientific method, the process generally follows a highly-structured and systematized framework of five main steps including direct observation, hypothesis formation, experimentation, data analysis, and conclusion drawing.

The scientific method begins with the direct observation of natural phenomena or the identification of a specific question or problem. Observations can arise from everyday experiences, previous research, or theoretical predictions. For example, a biologist might observe a pattern in animal behavior that prompts questions about its cause or function (West-Eberhard, 2024). This initial phase involves careful observation, documentation of relevant details, and formulating a clear research question or hypothesis that seeks to explain the observed phenomenon.

After forming a question, scientists develop a hypothesis—a testable explanation or prediction based on existing knowledge or theoretical frameworks (Mietchen et al., 2024). A hypothesis is a tentative statement that suggests a relationship between variables or offers an explanation for the observed phenomenon. Hypotheses are formulated to be specific, falsifiable (able to be tested and potentially proven false), and grounded in logical reasoning. For instance, based on observations, a physicist might hypothesize that a specific material will exhibit superconductivity at a certain temperature under controlled conditions (Chen, 2025).

The core of the scientific method involves designing and conducting experiments or empirically-driven investigations to test the hypothesis (Hirose et al., 2023). Experiments are carefully planned procedures that manipulate variables under controlled conditions to observe their effects and gather data. The experimental design aims to eliminate all forms of biases and confounding factors, ensuring that results are valid, reliable and reproducible. Data collected during experiments are analyzed using appropriate statistical methods to evaluate the validity of the hypothesis and draw meaningful conclusions. For example, in a psychology experiment, researchers might manipulate variables like environmental factors to observe their impact on human behavior and measure outcomes using standardized assessments (Haslam et al., 2024).

Upon collecting data, scientists analyze the results to determine whether they support or refute the hypothesis (Fife et al., 2024). Data analysis involves organizing, interpreting, and visualizing data using statistical tools and methods. This step aims to identify patterns, trends, and relationships within the data set. Statistical significance is often assessed to determine the likelihood that observed effects are due to chance. If the data support the hypothesis, scientists may revise or refine their hypothesis based on new insights gained from the results.

Through data analysis, scientists draw conclusions about the hypothesis and its implications (Dehalwar et al., 2024). These conclusions must be grounded in the empirically-driven and available evidence using critical thinking and logical reasoning to demonstrate how the findings advance existing scientific knowledge or theories. A key aspect of the scientific method is peer review—a thorough evaluation of research by independent experts in the field (Uttley et al., 2023). Expert peer review ensures that research findings are assessed for accuracy, methodological rigor, and ethical standards before being published in scientific journals or shared with the wider scientific community.

Real science is an iterative process where new discoveries and insights often lead to further questions and refinements of hypotheses (Perez-Bentancur, et al., 2024). If results do not support the hypothesis, scientists revise their approach, reconsider underlying assumptions, or propose alternative explanations for further investigation. This iterative cycle of observation, hypothesis testing, experimentation, and refinement drives scientific progress and contributes to the cumulative body of information and knowledge in a given field of study.

In summary, the scientific method of inquiry is a systematic and iterative process that guides scientific research and discovery. By emphasizing empirically-driven evidence, hypothesis testing, rigorous experimentation, data analysis, expert peer review, and continuous revision, the scientific method ensures that scientific conclusions are valid, reliable, reproducible, and open to enquiry. This highly-structured and organized approach underpins the enterprise of real science, enabling scientists to explore and understand the natural world, solve complex problems, and innovate for the benefit of society.

8.6 The Scientific Community's Role and Responsibility Under the Enterprise of Real Science.

The scientific community plays a pivotal role in advancing knowledge, fostering innovation, and addressing global challenges under the enterprise of real science (Ripple et al., 2024). Comprising of researchers, educators, institutions, and organizations, the scientific community operates within a framework of collaboration, expert peer review, and ethical responsibility to promote the integrity and impact of scientific research.

At its core, the scientific community is dedicated to expanding the frontiers of knowledge through rigorous inquiry and discovery (Jerome et al., 2024). Researchers conduct experiments, analyze data, and publish findings to contribute new insights and theories to their respective fields. Through continuous exploration and experimentation, scientists build upon existing knowledge, uncover new phenomena, and develop innovative solutions to complex and complicated problems in the natural world. This collective effort drives scientific progress, pushes the boundaries of human understanding, and lays the foundation for modern technological advancements that benefit society.

Collaboration is a hallmark of the scientific community (Amfo et al., 2024), promoting the exchange of ideas, resources, and expertise across disciplines, sectors, and institutions. Collaborative networks empower scientists to tackle complex challenges that demand diverse perspectives and specialized knowledge. Interdisciplinary research, which integrates insights from multiple sectors, fosters creativity and innovation by addressing problems from varied angles. Through collective efforts, scientists are better equipped to tackle pressing issues such as Anthropocene climate change, public health crises, and sustainable development than through isolated approaches.

Expert peer review is a cornerstone of scientific integrity (Uttley et al., 2023), ensuring that research findings undergo critical evaluation by independent subject matter experts before publication or dissemination. Peer-reviewed journals and conferences uphold demanding standards of methodological precision, ethical conduct, and theoretical relevance. This process helps to validate scientific claims, identify potential biases or errors, and maintain the credibility of scientific research within the broader community. Quality assurance mechanisms, such as reproducibility studies and data transparency initiatives, further enhance the reliability and strength of scientific findings.

The scientific community upholds ethical principles and guidelines to safeguard the welfare of research subjects, protect the environment, and promote transparency in scientific practices (Pols et al., 2024). Ethical considerations include informed consent for human and animal research, responsible use of research funds, and ethical conduct in data collection, analysis, and reporting. Scientists have a responsibility to communicate their findings accurately and engage with the public-at-large to promote understanding of scientific issues. Building public trust in real science through transparent communication and ethical behavior is foundational for fostering support for research funding, policy decisions, and evidence-based decision-making.

Education and mentorship are essential roles of the scientific community in nurturing the next generation of researchers and science enthusiasts (Suiter et al., 2024). Universities, research institutions, and professional societies provide training opportunities, workshops, and mentorship programs to support students and early-career scientists in developing advanced research skills, critical thinking abilities, logical reasoning, problem solving, and ethical practices. Mentors play a crucial role in guiding and inspiring young scientists just getting started by sharing their fund of knowledge and expertise alongside fostering a culture of scientific excellence and integrity.

The scientific community actively engages with policymakers, stakeholders, and the public to inform evidence-based decision-making and address societal challenges. Scientists contribute expertise to policy discussions on issues ranging from public health and biodiversity conservation to modern technological innovation and educational reform. By advocating for policies grounded in scientific evidence (Komalasari et al., 2024), the scientific community seeks to address global challenges, promote sustainable development, and improve the well-being of individuals and communities worldwide.

In summary, the scientific community plays multiple essential roles within the enterprise of real science. By advancing knowledge, promoting collaboration, upholding ethical standards, educating and training future scientists, and engaging with policymakers and the public, scientists contribute to the resilience, integrity, and impact of scientific research. Through collective effort and shared responsibility, the scientific community continues to drive progress, shape the future of real science, and address complex challenges facing society.

PART III

The Ethics of Real Science.

The ethics of real science encompasses the moral principles and established guidelines that govern scientific research, education, and practice, ensuring that the pursuit of information and knowledge is conducted transparently, responsibly and honestly. (Tomic et al., 2024). Key ethical considerations include honesty in reporting results, ensuring the welfare and rights of research subjects, and avoiding conflicts of interest that could compromise the validity of findings. Ethical real science requires transparency in methodology, respect for confidentiality, and the responsible use of data. Upholding these ethical standards is essential to maintaining public trust in real science, promoting reproducibility and credibility, and ensuring that scientific advancements contribute positively to society without causing harm.

CHAPTER 9

The Responsibilities of Scientists.

Scientists have significant responsibilities that extend beyond the pursuit of information and knowledge. They are entrusted with conducting research ethically, ensuring that their work upholds the highest standards of integrity and transparency. This includes accurately reporting data, acknowledging limitations, and avoiding any manipulation or falsification of results. Scientists are also responsible for considering the broader impact of their research on society, the environment, and future generations. This involves assessing potential risks, communicating findings effectively to the public, and engaging in dialogue about the ethical implications of their work. Additionally, scientists have a duty to educate, train, and mentor the next generation of research scientists, fostering a culture of competency, curiosity, critical thinking, transparency, accountability, and honesty.

Their dedication to these fundamental responsibilities not only advances the integrity and impact of scientific inquiry but also strengthens public trust in the scientific enterprise (Poincaré, 2022). By upholding rigorous methodologies, transparency, and ethical standards, scientists ensure that research remains credible, reproducible, and aligned with societal needs, preferences, and values. This commitment fosters greater public confidence in real science, promoting informed decision-making, empirically-driven evidence-based policy, and a deeper appreciation for the role of real science in addressing todays and tomorrow's global challenges.

9.1 The Professionalism of the Scientist.

Scientists are held to the highest level of trust and truthfulness by society. It was an expectation from the public-at-large during the COVID-19 pandemic that scientists would follow the "rules" of truthful real science and the methodology that goes with being dogmatic for the rules of observation and experimentation (Hamid, 2020).

Throughout history, extensive discourse has shaped the principles guiding how scientists approach their work within their respective fields. No one has written more authoritatively on the "new" morals and ethics of professionalism than the philosopher and historian Karl Popper. To begin, Popper created three main principles for any rational discussion between colleagues (Popper, 1998):

1. *The principle of fallibility:* Perhaps I am wrong and perhaps you are right; but, of course, we may both be wrong.
2. *The principle of rational discussion:* We need to test critically and, of course, as impersonally as possible the various theories that are in dispute.
3. *The principle of approximation to truth:* We can nearly always come closer to the truth with the help of such critical discussions; and we can nearly always improve our understanding, even in cases where we do not reach agreement.

These principles laid the foundation for a revised set of professional ethics essential for addressing contemporary scientific inquiry and methodology. Popper advocated for abandoning the outdated ethics rooted in personal knowledge and the pursuit of certainty—an approach that has contributed to one of the most problematic perspectives in modern science: the notion of absolute authority. Perone (2020) highlights that this traditional imperative, prevalent in both science and other professions, was built on the notion of *Be an authority! Know everything in your field of expertise!* According to Popper, this old ethical framework leaves no room for error, fosters intolerance, and, particularly in fields like medicine and politics, encourages the concealment of mistakes to preserve authority.

Perone (2020) aptly observed that during the COVID-19 pandemic, outdated ethical approaches became widespread. Many people unquestioningly deferred to the authority of epidemiologists rather than critically evaluating their statements, despite several errors made by experts, including inaccurate predictions from even Nobel laureates. This outdated ethical mindset also permeated politics, with numerous politicians opting to conceal their missteps rather than take responsibility for them.

In continuation of his work, Popper (1998) presented the scientific community with twelve principles of professional ethics, grounded in the distinction between objective knowledge and uncertain knowledge presented and paraphrased below:

1. The vast and ever-expanding body of objective knowledge in real science increasingly surpasses what any single individual can fully master. As a result, the concept of an absolute subject matter expert becomes untenable, even within highly specialized fields such as medicine.

2. It is impossible to completely avoid mistakes, even those that are theoretically preventable. All scientists make errors continually. The outdated notion that mistakes can always be avoided and that we have a duty to do so must be reconsidered, as this belief is, in itself, a mistake.

3. While we still have a responsibility to make every effort to avoid mistakes, we must acknowledge how difficult this task truly is. No one, not even the most intuitive and creative scientists, can avoid all errors. Although intuition is essential to scientific inquiry, it is more often wrong than right. Recognizing this helps us better navigate the challenges of minimizing mistakes.

4. Even in our most well-supported theories, mistakes may be lurking, and it is the scientist's responsibility to seek

them out. Uncovering errors in a well-corroborated theory or widely used practical method can lead to significant and valuable discoveries.

5.   We must shift our perspective on mistakes, as this is where ethical reform must start. The traditional professional ethic encourages us to hide our errors, keep them concealed, and quickly move on from them, which is a mindset we must change.

6.   The new guiding principle is that to minimize unnecessary mistakes, we must learn from the ones we do make. Concealing errors, therefore, becomes the greatest intellectual wrongdoing.

7.   We must always be vigilant in searching for mistakes, particularly our own. When we discover them, we should not forget them, but rather examine them from every angle to gain a deeper understanding of what went wrong.

8.   Adopting a self-critical mindset, along with honesty and openness toward oneself, becomes an essential part of everyone's responsibility.

9.   Since learning from our mistakes is essential, we must also graciously accept, and even appreciate, when others point them out. When we bring attention to someone else's errors, we should remember that we, too, have made similar mistakes, and even the greatest scientists have committed significant ones. This doesn't suggest that mistakes are always excusable; our vigilance must never waiver. However, recognizing that mistakes are inevitable can help us offer constructive feedback when addressing others' errors.

10.   We must recognize that we rely on others to help identify and correct some of our mistakes, just as they depend on us. This is particularly true for individuals who have been raised with different perspectives and in distinct cultural environments. Such interactions should occur with cul-

tural humility and empathy while fostering tolerance and understanding.

11. We must recognize that self-criticism is the most valuable type of critique, but feedback from others is also foundational and capacity building. Although it may not be as impactful as self-reflection, it remains extremely valuable.

12. Rational (or objective) criticism must always be specific, providing clear reasons for why particular statements, hypotheses, or arguments seem false or invalid. It should be aimed at moving closer to objective truth. In this regard, criticism should be both impersonal and empathetic.

Perone (2020) opines:

"... by accepting our conjectural knowledge, we naturally end up devising ethics for tolerance, for self-criticism, intellectual honesty, and the much-needed criticism from others, especially from different cultural atmosphere, as it is upon divergence that we grow knowledge."

9.2 Scientists, Society, and the Environment.

Scientists are responsible for considering the broader impact of their research on society, the environment, and future generations (Leichenko et al., 2024). This responsibility requires them to look beyond the immediate results of their work and evaluate how their findings could affect various aspects of life, including social structures, environmental health, and the well-being of future populations. Whether developing new technologies, studying ecosystems, or exploring medical advancements, scientists must be mindful of the potential long-term consequences of their research. This foresight helps ensure that scientific progress does not come at the expense of ethical standards or the greater good.

Assessing potential risks is a critical component of this responsibility (Williams et al., 2024). Scientists must anticipate and evaluate the possible dangers their work might pose, whether to human health, environmental stability, or societal norms. This risk assessment often involves interdisciplinary and multisectoral collaboration, where scientists work with ethicists, policymakers, and other stakeholders to thoroughly examine the implications of their findings. By doing so, they can develop strategies to mitigate negative outcomes, ensuring that the benefits of their research are maximized while minimizing harm.

Communicating findings effectively to the public is another key aspect of a scientist's responsibility (Szczuka et al., 2024). Clear, accurate, and accessible communication helps bridge the gap between complex scientific concepts and public understanding. It allows society to make informed decisions about the adoption of advanced technologies, policies, or practices based on empirically-driven evidence. Scientists must be transparent about their methods, results, and any uncertainties in their work, helping to build trust with the public-at-large and prevent the spread of any misinformation. Engaging with the public also provides an opportunity for scientists to listen to societal concerns and integrate them into their research, creating a more inclusive and responsive scientific community.

Finally, engaging in dialogue about the ethical implications of their work is essential for scientists (Swain et al., 2020). This involves not only reflecting on the moral aspects of their research but also actively participating in discussions with peers, ethicists, and the public. Such dialogue ensures that scientific progress aligns with societal values and ethical principles. It also promotes accountability, as scientists are called upon to justify their work and its impact. By embracing this responsibility, scientists contribute to a culture of ethical awareness and responsibility that guides the direction of scientific discovery for the benefit of all.

9.3 The Scientific Community.

The scientific community serves a multi-layered purpose, acting as the collective engine that drives the advancement of knowledge and understanding in the natural world. At its core, the scientific community exists to foster collaboration among researchers, providing a safe platform where ideas can be shared, tested, and refined. This collaborative environment is essential for solving complex problems that require expertise from multiple disciplines. For example, addressing climate change necessitates the combined efforts of climatologists, biologists, chemists, economists, and policymakers (Molthan-Hill et al., 2022). By working together, the scientific community can approach these challenges from various angles, leading to more comprehensive and effective solutions.

Another critical purpose of the scientific community is to uphold the integrity of the research process. Through expert peer review, scientists critically evaluate each other's work before it is published, ensuring that the research meets the necessary standards of quality, accuracy, reliability, and reproducibility. This process acts as a safeguard against errors, biases, and fraudulent practices, maintaining the credibility of scientific findings. For instance, a new medical treatment must undergo rigorous subject matter expert peer review and highly-regulated clinical trials before it can be approved for public use. This scrutiny helps protect the publics' health and safety by ensuring that only thoroughly vetted and validated therapies reach patients (Krychtiuk et al., 2024).

The scientific community also plays a vital role in education, training, and mentorship (Deng et al., 2024). Experienced scientists are responsible for guiding the next generation of scientists, imparting knowledge, skills, and ethical principles that are foundational and capacity-building for their development. This mentorship helps maintain a continuous flow of innovation and discovery, as young scientists are encouraged to ask new questions, explore novel ideas, and push the boundaries of what is known. For example, a seasoned physicist might mentor a graduate student, helping them navigate the complexities of quantum mechanics and encouraging them to pursue groundbreaking research in the field.

Furthermore, the scientific community serves as a bridge between real science and society (Galdames et al., 2024; Beuchert et al., 2024). Scientists have a responsibility to communicate their findings to the public in a way that is clear, accurate, and accessible. This communication helps ensure that the public is informed about important scientific developments and can make decisions based on empirically-derived evidence. For instance, during the COVID-19 pandemic, the scientific community played a crucial role in educating the public about the virus, vaccines, and protective measures (Wie et al., 2023; Intemann, 2023). By doing so, they helped guide public behavior and public policy decisions that were essential in managing the crisis.

Finally, the scientific community is responsible for setting the ethical standards that guide research practices (Swain et al., 2020). This includes addressing the moral implications of scientific work, such as the potential impact of genetic engineering (Nicholl, 2023), artificial intelligence (Osasona et al., 2024), or environmental interventions (Hourdequin, 2024). By engaging in ongoing ethical discussions, the scientific community ensures that research not only advances knowledge but also aligns with societal needs, preferences, and values and contributes positively to the common good. This ethical oversight is essential for maintaining public trust in real science and ensuring that scientific progress benefits humanity as a whole.

CHAPTER 10

The Intersection of Real Science and Society.

The intersection of real science and society is a dynamic space where knowledge, ethics, and cultural needs, preferences, and values converge to shape human progress (Leichenko et al., 2024). In this space, scientific discoveries influence societal norms, public policies, and practices, driving advancements in medicine, technology, and resource stewardship. At the same time, societal values guide the direction of scientific inquiry, setting ethical boundaries and prioritizing research that addresses pressing global challenges. This interplay ensures that real science not only advances knowledge but also serves the broader interests of humanity, fostering a world that is both informed by empirically-derived evidence and grounded in cultural humility and empathy.

10.1 How Real Science Influences and is Influenced by Societal Values.

Real science and societal values are inextricably interconnected, each shaping and influencing the other in significant ways. Real science often reflects the values, priorities, and concerns of the society in which it is conducted. For example, research funding and focus areas are frequently guided by what society deems important at a given time. During periods of health crises, such as the COVID-19 pandemic (Butterworth et al., 2024), society placed a high value on medical research, leading to rapid advancements in vaccines and treatments. Conversely, in times of peace and prosperity, scientific exploration might prioritize other areas, such as space exploration (Lin et al., 2024) or environmental stewardship (Kumar, 2024), reflecting society's broader ambitions and ethical considerations.

Societal values also play a crucial role in determining the ethical boundaries of scientific research. What one society views as acceptable or desirable in scientific endeavors may differ significantly from another. For instance, debates around genetic modification, cloning, or stem cell research are heavily influenced by cultural, religious, and ethical values (von Schwarz, 2024 a). These societal perspectives can either propel scientific innovation or impose restrictions to ensure that research aligns with prevailing moral standards. This dynamic ensures that real science progresses within a framework that is sensitive to the diverse values held by different communities, ultimately guiding the direction and application of scientific discoveries.

On the other hand, real science has the power to challenge and transform societal values. Scientific discoveries often prompt shifts in how society understands and interacts with the world. For example, the theory of evolution by natural selection (Darwin, 1859), revolutionized the way people perceived the origins of life, challenging traditional views and prompting profound changes in education, religion, and philosophy. Similarly, advancements in understanding human psychology and neuroscience (Collins, 2023), have influenced societal views on mental health, leading to more compassionate approaches to healthcare service delivery and reducing societal stigma. In this way, real science not only responds to societal values but also acts as a catalyst for cultural and ethical evolution.

Real science can also influence societal values by providing evidence that informs public policy and ethical debates. For example, climate science has significantly shaped global environmental policies and has fostered a growing societal commitment to stewardship and conservation (Koonin, 2024). Scientific evidence highlighting the impacts of pollution, deforestation, and greenhouse gas emissions has led to a shift in public attitudes and a greater emphasis on ecological responsibility (Kaiho, 2023). These changes in societal values often translate into actionable policies and practices, such as the adoption of renewable energy sources (Adelekan et al., 2024) or the implementation of conservation efforts (Langhammer et al., 2024), illustrating how real science can drive societal change.

The relationship between real science and societal values is both dynamic and reciprocal. Societal values influence the direction and ethical framework of scientific inquiry, while real science, in turn, reshapes those values by advancing knowledge, challenging established beliefs, and offering fresh empirically-driven evidence-based perspectives. This continuous exchange keeps real science relevant and responsive to society's needs, preferences, and aspirations, while simultaneously driving progress and expanding the limits of what is possible, often paving the way for a more informed and forward-thinking world.

10.2 The Ethics of Controversial Research Areas.

The ethics of controversial research areas today is a complex and evolving landscape, reflecting the tension between scientific innovation and societal values.

One of the most debated areas is genetic modification, particularly in humans (Rueda, 2024). The development of CRISPR technology has enabled gene editing (Feng et al., 2024) with remarkable precision, offering the potential to cure genetic diseases or even enhance human capabilities. However, this promise is accompanied by serious ethical concerns. Critics fear the rise of *designer babies* (Capalbo et al., 2024), where genetic enhancements could create new forms of inequality or result in unforeseen consequences that may not become apparent until it's too late. Ethical debates surrounding this advanced medical technology often focus on issues of consent, the risk of unintended harm, and its broader societal impact. The challenge is to balance the drive for transformational medical break-throughs with the responsibility to prevent misuse and ensure fair access for all.

Another controversial area is artificial intelligence (AI), particularly in its application to autonomous systems and surveillance (So, 2024). AI has the potential to revolutionize industries, improve healthcare, and enhance everyday life, but it also poses significant ethical dilemmas. For instance, the development of autonomous weapons raises profound moral questions about accountability in warfare, while AI-driven surveillance technologies challenge notions of privacy and civil liberties. The ethical concerns here include bias in AI algorithms, the potential for mass surveillance to erode personal freedoms, and the risks of creating systems that could operate beyond human control. As AI continues to advance (Sherani et al., 2024), it is crucial to establish ethical guidelines that protect individual rights and ensure that these technologies are used in ways that benefit society as a whole.

Stem cell research, particularly the use of embryonic stem cells, remains a contentious ethical issue (Park et al., 2024). Stem cells offer the potential to treat or even cure conditions like Parkinson's disease, diabetes, and spinal cord injuries, but the use of embryonic stem cells raises ethical concerns about the moral status of embryos. Critics argue that using embryos in research is unethical, likening it to the destruction of potential human life. Supporters, on the other hand, highlight the potential for groundbreaking medical advances and contend that the research is justified if it leads to significant progress in treating debilitating diseases. The ethical debate centers on finding a balance between the promise of medical breakthroughs and the moral considerations tied to the source of the stem cells.

Climate engineering, or geoengineering, is another area fraught with ethical complexities (Symons et al., 2024). As the impacts of man-made climate change become increasingly severe, some scientists are exploring large-scale interventions to artificially alter the Earth's climate, such as solar radiation management or carbon capture and storage. While these technologies could potentially mitigate the effects of man-made climate change, they also come with significant risks and ethical questions. Critics argue that geoengineering could have unforeseen and potentially catastrophic consequences for the environment and could be used as a justification to delay more sustainable solutions to climate change. The ethics of climate engineering involve considerations of environmental justice, the potential for unintended harm, and the moral responsibility to protect future generations.

Lastly, research involving animal testing continues to be a controversial ethical issue (von Schwarz, 2024b). Animal research has played a key role in numerous medical breakthroughs, yet it raises important concerns about the welfare and rights of animals. Ethical debates often center on the necessity of using animals in research, the conditions in which they are kept, and the potential for their suffering. A growing movement advocates for reducing, refining, and replacing animal testing with alternative methods, but the challenge lies in balancing scientific advancement with the ethical treatment of animals. The ethics of animal testing demand careful reflection on both the scientific gains and the moral implications of using live animals in research.

10.3 Real Science in Healthy Public Policy: Balancing Scientific Evidence with Ethical Considerations.

Scientific research is vital in shaping comprehensive healthy public policy (Straf et al., 2012) by providing empirically driven, evidence-based insights that guide fair and informed decisions on a broad spectrum of issues, from essential public health services to environmental protection to food and housing security (Pallett, 2020). However, incorporating scientific evidence into healthy public policy requires thoughtful consideration of ethical dimensions (Namdarian et al., 2024). While data-driven approaches are foundational, they must be interpreted within the broader context of societal values, cultural norms, and moral principles. For example, even if scientific evidence supports the effectiveness of public health interventions like mandatory vaccinations (Goldsteen et al., 2024), policymakers must also account for ethical concerns such as individual autonomy and informed consent (Pugh et al., 2024). Balancing these factors ensures that healthy public policies are not only effective but also ethically sound, aligning financial incentives with the values of the communities they serve.

One of the challenges in balancing scientific evidence with ethical considerations is the potential for conflicts between scientific recommendations and public sentiment. For instance, climate change policy is often grounded in robust scientific evidence (Koonin, 2024) that underscores the need for urgent action to reduce greenhouse gas emissions. However, implementing such policies can raise ethical concerns related to economic impacts, particularly for communities that depend on industries tied to fossil fuels. Policymakers must navigate these complexities by considering not only the scientific imperative to protect the environment but also the ethical obligation to support vulnerable and marginalized populations during the transition to a more sustainable economy. This requires a nuanced approach that integrates scientific insights with strategies for equitable and social justice (Ashcraft et al., 2024).

Balancing real science and ethics is equally critical in regulating emerging and advanced medical technologies like artificial intelligence and biotechnology. While scientific advancements in these areas promise significant benefits, they also raise complex ethical concerns, such as privacy, cybersecurity, and human rights. For instance, the use of AI in law enforcement (Pandey et al., 2024) may enhance crime detection and prevention, but it also introduces risks related to institutional bias, explicit racism, excessive surveillance, and the erosion of individual's civil liberties. Policymakers must carefully weigh the potential scientific gains against these ethical challenges, crafting regulations that safeguard individual rights while promoting innovation and creativity. Striking this balance is vital for maintaining public trust in real science, technology, and government.

Healthy public policy provides another example of the intersection between real science and ethics. During the COVID-19 pandemic (Greer et al., 2024), policymakers relied heavily on scientific evidence to guide decisions about quarantine, isolation, social distancing, and vaccine distribution. However, these measures also had significant ethical implications, such as the impact on individual freedoms, economic livelihoods, and behavioral health. Policymakers faced the challenge of balancing the need to protect the public commons with the ethical obligation to minimize harm to individuals, families, neighborhoods, and communities. This required ongoing dialogue between scientists, ethicists, and the public-at-large to ensure that healthy public policies were empirically-driven, evidence-based, and ethically sound.

In conclusion, real science in healthy public policy demands a careful balance between empirically-derived evidence and ethical considerations. While real science provides the data and insights necessary to address complex societal challenges, ethical considerations ensure that healthy public policies are implemented in ways that respect human rights, promote equity, and align with societal values. This balance is essential for creating policies that are not only effective and sustainable but also just and humane, reflecting the needs and values of the communities they are designed to serve.

CHAPTER 11

The Role of Real Science in Education.

Real science plays a pivotal role in education by fostering critical thinking, problem-solving, and logical reasoning skills while simultaneously investigating a deeper understanding of the natural world. Integrating scientific principles into the curriculum encourages students to explore and question how phenomena occur, promoting a hands-on approach to learning through experiments and inquiry-based activities. This engagement with real science not only enhances students' knowledge of fundamental concepts but also builds capacity for them in pursuing future careers in STEM (science, technology, engineering, and mathematics) fields (Moomaw, 2024). By cultivating curiosity and a scientific mindset, education equips students with the tools, information, and knowledge to address complex challenges, make informed and rational decisions, and contribute meaningfully to society.

11.1 Overarching Framework for Scientific Literacy.

Scientific literacy refers to an individual's understanding of scientific concepts and processes necessary for personal decision-making, participation in civic and cultural affairs, and economic productivity (Rudolph, 2024). The overarching framework of scientific literacy includes several key components that establishes the evaluation of success vs. failure of the enterprise of real science.

A foundational understanding of concepts from scientific disciplines (e.g., biology, chemistry, physics, and earth sciences) is essential for developing scientific literacy. This includes grasping the nature of real science, recognizing real science as a method of discovery, understanding the scientific method, and appreciating the iterative and evolving nature of scientific knowledge. Scientific inquiry, critical thinking, problem-solving, and logical reasoning are fundamental skills for asking questions, designing experiments, collecting and analyzing data, and drawing conclusions based on evidence. Applying scientific concepts and knowledge to real-world situations, particularly in addressing societal and environmental challenges, is also key. Individuals should be able to critically assess scientific information, differentiate between empirically-driven, evidence-based conclusions and personal opinions, and make informed decisions. Additionally, communicating scientific ideas effectively through writing, speaking, and visual representation is important. Understanding the impact of real science and technology on the modern world, including their ethical, legal, and social implications, is critical. Lastly, fostering a positive attitude towards real science involves nurturing curiosity, openness to new ideas, skepticism of unsupported claims, and an appreciation for real science's contributions to society.

These components work together to equip individuals with the tools, information, and knowledge they need to understand and engage with scientific issues, make informed decisions, and participate fully in a real science-driven world.

11.2 Why Teach Real Science?

Teaching students about real science is essential for fostering a well-rounded education and preparing individuals to navigate and contribute to an increasingly complex scientific and technologically-driven world (Almasri, 2024). Real science education cultivates scientific inquiry, critical thinking, problem-solving, logical reasoning, and a curiosity about the natural world, which are invaluable assets in both personal and professional contexts. By understanding scientific principles and methods, students can make informed decisions, assess the truthfulness of information, and engage with contemporary issues that collectively impact society.

One of the primary reasons to teach science is to develop a scientifically literate populace (Rudolph, 2024). Scientific literacy enables individuals to understand and evaluate scientific information that they encounter in everyday life, from health and nutrition advice to environmental conservation and technological advancements. In an era where disinformation, misinformation and pseudoscience can easily spread, the ability to critically assess sources and claims is crucial. A solid grounding in real science helps individuals discern sound information, understand the empirically-derived evidence behind it, and make rational, informed choices that affect their health, safety, and well-being.

Real science education fosters curiosity and a sense of wonder about the natural world. Through the exploration of scientific concepts and participation in hands-on experiments, students learn to ask questions, seek evidence, and develop a deep appreciation for the complexities of the universe. This curiosity not only enhances their educational journey but also promotes a lifelong passion for learning and discovery. Moreover, sparking an interest in science at an early age can inspire future careers in STEM fields, which are essential for driving innovation and supporting economic growth for the next generation of scientists (Moomaw, 2024).

Moreover, teaching science equips students with valuable skills that are transferable to various aspects of life and work. The scientific method—observation, hypothesis formulation, experimentation, and analysis—instills a disciplined approach to problem-solving that can be applied beyond scientific contexts. Students learn to think logically, analyze data, draw conclusions based on evidence, and communicate their findings effectively. These hard skills are highly sought after in many professions, from engineering and healthcare to business and public policy, making real science education a vital component of career readiness.

In addition, real science education plays a crucial role in addressing global challenges (Gustian et al., 2024). Issues such as man-made climate change, public health crises, and sustainable societal development require both a scientifically-informed citizenry and an educated assemblage of policymakers. By understanding the scientific underpinnings of these challenges, students are better equipped to engage in meaningful discourse, advocate for empirically-driven evidence-based policies, and contribute to solutions that benefit society and the planet. Real science education thus fosters a sense of responsibility and empowerment, enabling individuals to actively engage in addressing the pressing interdisciplinary and multisectoral problems of our time.

In conclusion, teaching real science is fundamental to fostering a scientifically literate society, where individuals can critically evaluate information, engage in evidence-based reasoning, and apply logical problem-solving skills to real-world issues. A robust real science education nurtures curiosity and encourages lifelong learning, enabling students to ask meaningful questions, challenge assumptions, and develop innovative solutions to complex problems.

Beyond individual benefits, real science education plays a crucial role in addressing global challenges such as climate change, public health crises, and technological advancements. By equipping students with the ability to analyze data, conduct experiments, and interpret scientific findings, we prepare them to contribute to a world that increasingly relies on scientific knowledge to inform policy decisions, drive economic growth, and improve quality of life.

Moreover, fostering a deep understanding of real science empowers future generations to pursue careers in STEM fields, where they can push the boundaries of discovery and innovation. It also strengthens democratic societies by enabling citizens to make informed decisions on critical issues, from healthcare choices to environmental policies.

Ultimately, investing in real science education is an investment in progress. It ensures that individuals not only understand the world around them but also actively participate in shaping a future driven by reason, knowledge, and discovery. By prioritizing science literacy, we lay the foundation for a more informed, innovative, and resilient society for generations to come.

11.3 The Next Generation Science Standards

The Next Generation Science Standards (NGSS) (NGSS Lead States, 2013) establish a comprehensive framework for real science education that emphasizes the integration of scientific knowledge with the practices that scientists and engineers use in their work. These standards are designed to prepare students not only to understand scientific concepts but also to engage in scientific inquiry and engineering design. The NGSS outlines eight (8) Science and Engineering Practices (SEPs) that are central to fostering a deeper understanding of real science and its application to real world problems.

One foundational practice is asking questions (for real science) and defining problems (for engineering). This practice involves students in the initial steps of scientific inquiry and engineering design, encouraging them to be curious, pose meaningful questions, and identify real-world problems. By asking questions and defining problems, students learn to think critically and to consider the scope, scale, and implications of scientific investigations and engineering projects (Jonassen et al., 2015). This practice lays the groundwork for further exploration and innovation.

Another essential practice is developing and using models. In real science, models are used to represent phenomena and to predict outcomes, while in engineering, they help in the design and testing of solutions. Students engage in creating, testing, and refining models to expand their understanding of scientific concepts and to visualize complex systems. This practice enhances students' ability to abstract and simplify reality, which is crucial for both scientific discovery and engineering problem-solving (Alismail et al., 2015).

Planning and carrying out investigations are a practice that directly involves students in the process of scientific experimentation and data collection. It encourages them to design and conduct their own experiments, to gather and analyze data systematically, and to refine their methods based on their findings. This hands-on approach helps students to develop a deeper appreciation for the empirical nature of real science and to gain practical skills in conducting rigorous scientific research (Pajo, 2022).

Analyzing and interpreting data is another foundational practice that involves examining data to identify patterns, draw conclusions, and make empirically-driven evidence-based decisions. This practice teaches students to handle data critically and to use statistical and graphical tools to make sense of their findings (Brookfield, 2011). It also emphasizes the importance of data integrity and the need to question and verify results (Jain et al., 2024), which are essential skills for both scientists and engineers.

Using mathematics and computational thinking is integral to modern real science and engineering today. This practice encourages students to apply mathematical concepts and computational techniques to solve problems, model phenomena, and analyze large amounts of data. By integrating mathematics and computational skills, students enhance their ability to quantify and rigorously analyze scientific and engineering problems (Krakowski et al., 2024), which is critical for innovation and technological advancement.

Constructing explanations (for real science) and designing solutions (for engineering) is a practice that synthesizes students' understanding and creativity. In real science, constructing explanations involves using empirically-derived evidence to build coherent and logical accounts of natural phenomena (Alvesson et al., 2024). In engineering, designing solutions requires applying scientific principles to create viable and effective solutions to real-world problems (Buede et al., 2024). This practice emphasizes the iterative nature of both scientific inquiry and engineering design, where ideas are continually tested and refined.

Engaging in rational discourse driven from empirically-derived evidence involves evaluating and defending scientific explanations or engineering solutions. Rational discourse fosters critical thinking and communication skills, as students learn to present and critique arguments, consider alternative explanations, and justify their conclusions with empirical data (Godfrey, 2023). It mirrors the collaborative and evaluative processes used by scientists and engineers to advance knowledge and technology.

Finally, obtaining, evaluating, and communicating information is a practice that encompasses the ability to gather information from diverse sources, critically assess its truthfulness and reliability, and effectively communicate findings within and outside of the scientific community. This practice is essential for scientific literacy, as it enables students to engage with scientific literature, collaborate with peers, and share their discoveries with broader audiences (Osbourne, 2023).

In conclusion, the Next Generation Science Standards (NGSS) Science and Engineering Practices (SEPs) provide a comprehensive and dynamic framework that bridges scientific knowledge with the practical skills and methodologies used by scientists and engineers. These practices go beyond rote memorization of facts by actively engaging students in the processes of scientific inquiry and engineering design, fostering a deeper and more meaningful understanding of how real science and technology shape the world around us.

By incorporating real-world applications and hands-on learning experiences, the NGSS SEPs cultivate essential skills such as critical thinking, problem-solving, and logical reasoning. Students learn to ask questions, develop and test hypotheses, analyze data, construct explanations, and design innovative solutions—mirroring the processes used by professionals in scientific and engineering fields. This inquiry-based approach enhances comprehension, strengthens cognitive abilities, and promotes adaptability, preparing students to tackle complex challenges with confidence and creativity.

Furthermore, the integration of real science and engineering practices equips students with the tools to become informed, responsible, and engaged citizens in an increasingly advanced technological and data-driven society. With rapid advancements in artificial intelligence, biotechnology, environmental stewardship, and other STEM-related fields, it is crucial that students develop the ability to critically assess scientific claims, participate in evidence-based decision-making, and contribute meaningfully to societal progress.

Ultimately, the NGSS Science and Engineering Practices represent a transformational approach to real science education, emphasizing not only the acquisition of knowledge but also its practical application in problem-solving and innovation. By fostering curiosity, resilience, and a commitment to discovery, these practices ensure that future generations are well-equipped to navigate and shape the scientific and technological landscapes of the 21st century.

11.4 What are the Cross-cutting Concepts Established by the NGSS?

The Next Generation Science Standards (NGSS) include Crosscutting Concepts (CCCs) (NGSS Lead States, 2013) as one of the three dimensions of real science education alongside Disciplinary Core Ideas and Science and Engineering Practices. These concepts are designed to help students develop a coherent and scientifically-based view of the world by making connections across different scientific disciplines. The seven (7) NGSS Crosscutting Concepts (CCCs) provide a framework for students to understand and apply overarching ideas that are relevant to various fields of real science.

Patterns is the first crosscutting concept, emphasizing the identification and analysis of recurring shapes, structures, and events in nature. Recognizing patterns allows students to make predictions and identify relationships between different phenomena. For example, observing the regular phases of the moon helps students understand lunar cycles and predict future phases (Kavanaugh et al., 2005). By studying patterns, students can develop a deeper understanding of natural laws and principles that govern the universe.

Cause and effect or mechanism and explanation focuses on understanding how and why things happen. This concept helps students to explore causal relationships and mechanisms behind observed phenomena. By investigating cause-and-effect relationships, students learn to develop explanations and predict outcomes. For example, understanding the cause-and-effect relationship between greenhouse gas emissions and man-made climate change enables students to grasp the mechanisms driving global warming and its impacts (Kirshbaum, 2024).

Scale, proportion, and quantity involves the understanding of size, time, and energy scales in scientific phenomena. This concept helps students appreciate the relative magnitudes and durations of different processes and objects, from atomic particles to galaxies, and from nanoseconds to geological epochs. Understanding scale, proportion, and quantity allows students to comprehend the significance of scientific measurements and the relationships between different quantities in scientific equations and models (Kapon et al., 2024).

Systems and system models emphasize the concept of a system as a set of interacting components working together. This concept encourages students to think about the components and processes within a system, how they interact, and how they influence the system as a whole. For example, studying an ecosystem as a system of interacting organisms and their physical environment helps students understand the complex interdependencies and the impact of changes within the system (Dozier et al., 2024; Little et al., 2023). System models are used to simplify and represent these complex interactions, aiding in analysis and prediction.

Energy and matter, flows, cycles, and conservation focus on the movement, transformation, and conservation of energy and matter in natural and human-made complex systems. This concept helps students understand that energy and matter cannot be created or destroyed but can change forms and move through systems in cycles.

For example, the water cycle and the carbon cycle illustrate the continuous movement and transformation of matter on Earth (Reichle, 2023). Understanding these flows and cycles is crucial for grasping how complex systems maintain balance and how human activities can disrupt them.

Structure and function explore the relationships between the shape and structure of an object or system and its function or behavior. This concept helps students understand that the way something is built or organized determines how it works. For instance, the structure of a bird's wings is directly related to their function in flight (Rader et al., 2023). By analyzing structure and function, students can develop insights into the design and operation of both natural and engineered systems.

Stability and change examine the conditions under which systems are stable or change over time. This concept helps students understand that systems can be in a state of equilibrium or undergo significant changes due to internal or external influences. For example, understanding the factors that contribute to the stability of ecosystems or the conditions that lead to their collapse is crucial for environmental science (Keith et al., 2023). Recognizing the dynamics of stability and change enables students to anticipate and respond to changes in natural and engineered systems.

In summary, the Crosscutting Concepts (CCCs) established by the Next Generation science Standards (NGSS Lead States, 2013) provide students with a set of lenses through which to view and understand the natural and technological world. By emphasizing patterns, cause and effect, scale, systems, energy and matter, structure and function, and stability and change, these crosscutting concepts help students make connections across different scientific disciplines, develop a holistic understanding of scientific principles, and apply their knowledge to solve complex problems. These overarching ideas are integral to fostering scientific literacy and preparing students to think critically and logically about the natural world around them.

11.5 Disciplinary Core Ideas Established by the NGSS.

The Disciplinary Core Ideas (DCIs) established by the Next Generation Science Standards (NGSS) form a crucial component of the three-dimensional framework designed to enhance real science education. DCIs represent the fundamental concepts and principles that students need to understand across various scientific disciplines. They are organized into four primary domains: *Physical Sciences, Life Sciences, Earth and Space Sciences, and Engineering, Technology, and Applications of Science.* These core ideas provide a coherent and cumulative foundation of knowledge that students build upon as they progress through their education.

In the *Physical Sciences*, the DCIs focus on matter and its interactions, motion and stability, energy, and waves and their applications. These ideas encompass fundamental principles such as the structure and properties of matter, chemical reactions, the laws of motion, the principles of energy conservation and transfer, and the nature of electromagnetic radiation. For example, understanding how energy is conserved and transferred in various processes enables students to grasp concepts from chemical reactions to thermal dynamics and electrical circuits (Atkins et al., 2023; Kuhn et al., 2023). These core ideas provide students with a robust understanding of the physical principles that govern the natural world.

Life Sciences DCIs cover the structures and processes of living organisms, ecosystems, heredity, and biological evolution. These ideas delve into the organization and function of cells, the mechanisms of genetics and inheritance, the diversity and interdependence of ecosystems, and the evidence and processes of evolution. For instance, learning about the genetic basis of traits and how they are passed from one generation to the next helps students understand the foundations of heredity and evolution (Hanisch et al., 2023). These core ideas enable students to explore the complexity of living systems and the interactions that sustain life on Earth.

In the *Earth and Space Sciences*, the DCIs encompass Earth's systems, Earth and human activity, and the universe and its stars. These ideas include the study of the Earth's materials and processes, the water cycle, weather and climate, natural resources, and the impact of human activities on the environment. They also cover the broader concepts of the solar system, the formation and evolution of stars, and the dynamics of galaxies. For example, understanding the water cycle and its impact on weather and climate patterns equips students with the knowledge to analyze and respond to environmental changes and challenges (Mercan, 2024). These core ideas help students appreciate the interconnectedness of Earth's systems and our place in the universe.

The *Engineering, Technology, and Applications of Science* DCIs emphasize the principles and practices of engineering and the use of technology to solve real-world problems. These ideas focus on the engineering design process, the links between engineering and scientific inquiry, and the applications of scientific knowledge to develop new technologies. For example, learning about the engineering design process—from defining problems to designing, testing, and optimizing solutions—prepares students to think creatively and critically about technological challenges (Buede et al., 2024). These core ideas bridge the gap between scientific principles and their practical applications, fostering innovation and problem-solving skills.

The integration of the Disciplinary Core Ideas with Science and Engineering Practices and Crosscutting Concepts ensures that students not only acquire knowledge but also understand how to apply it in various contexts. By exploring these core ideas through hands-on investigations, modeling, and problem-solving activities, students develop a deep and meaningful understanding of real science that goes beyond memorization. They learn to connect concepts across different domains, apply scientific reasoning to real-world issues, and appreciate the relevance of science and engineering in their everyday lives.

In summary, the Disciplinary Core Ideas established by the Next Generation Science Standards (NGSS Lead States, 2013) provide a structured and comprehensive foundation for real science education. Covering the *Physical Sciences, Life Sciences, Earth and Space Sciences, and Engineering, Technology, and Applications of Science*, these core ideas equip students with the essential knowledge and technical skills needed to understand and engage with the natural and technological world. By integrating these core ideas with scientific practices and crosscutting concepts, the NGSS framework promotes a holistic and inquiry-based approach to real science education, preparing students to become informed and capable problem-solvers in an increasingly complex and interconnected world.

11.6 Basic Understandings About the Nature of Science.

The National Science Teaching Association (NSTA) in 2021 established a set of basic understandings about the Nature of Science (NOS) to provide a clearer framework for real science education (NSTA, 2021). These understandings emphasize that real science is not just a body of knowledge but a dynamic process of inquiry and discovery. They aim to foster scientific literacy by helping students grasp the fundamental characteristics and processes that define scientific endeavors. Understanding NOS helps students appreciate how scientific knowledge is developed, validated, and applied, providing a foundation for critical thinking and informed decision-making.

One of the key understandings is that real science is a way of knowing. This principle highlights that real science seeks to build reliable knowledge about the natural world through empirically-derived evidence and logical reasoning. Unlike other ways of knowing, such as philosophy or religion, real science relies on observable, testable, and repeatable evidence. Empirically-driven evidence allows scientific knowledge to be continually tested, refined, and expanded. Teaching students about this aspect of real science helps them understand the importance of empirically-derived evidence and the methods scientists use to investigate and understand natural phenomena.

Another crucial aspect is that scientific knowledge is subject to change. Real science is inherently provisional, meaning that scientific explanations and theories can be revised or replaced as new evidence emerges. This dynamic nature of real science underscores the importance of skepticism, open-mindedness, and critical evaluation of new data. Students learn that scientific knowledge is not absolute but evolves over time through the process of investigation, expert peer review, and replication. This understanding fosters an appreciation for the ongoing advancement of real science and the role of innovation in driving scientific progress.

The NSTA (2021) also emphasizes that real science is a human endeavor. This understanding recognizes that real science is conducted by individuals and communities, influenced by their creativity, curiosity, and cultural contexts. It acknowledges that scientists are driven by a desire to understand the natural world and solve problems, and that their work is shaped by societal needs, preferences, and values. By exploring the human aspects of real science, students can appreciate the diversity of people involved in scientific research, the collaborative nature of scientific work, and the ethical considerations that guide scientific practice.

Additionally, the NSTA (2021) outlines that real science and engineering are inextricably interconnected. While real science seeks to understand natural phenomena, engineering applies scientific principles to design and build solutions to practical problems. This understanding highlights the symbiotic relationship between real science and engineering, where advances in one field often lead to progress in the other. Students learn that the engineering design process involves defining problems, developing solutions, and iterating designs based on testing and feedback. This integration helps students see the relevance of real science and engineering in addressing real-world challenges and fosters skills in problem-solving and innovation.

Lastly, the NSTA (2021) points out that scientific knowledge relies on a variety of scientific methods. There is no single "scientific method" but rather a diverse set of approaches that scientists use to gather data, test hypotheses, and develop theories. These methods include experimentation, observation, modeling, and comparative analysis, among others. Understanding the variety of methods used in real science helps students appreciate the flexibility and creativity involved in scientific research. It also underscores the importance of using appropriate methods for different types of questions and investigations.

In summary, the basic understandings about the Nature of Science established by the National Science Teaching Association in 2021 emphasize that real science is an empirical, dynamic, and human endeavor inextricably interconnected with engineering and reliant on diverse methods. These understandings aim to deepen students' appreciation of real science as a way of knowing, foster critical thinking, and prepare them to engage thoughtfully with scientific issues. By integrating these concepts into real science education, the NSTA (2021) seeks to cultivate a scientifically literate society equipped to navigate and contribute to an increasingly complex world.

Below is a summary from NSTA's Nature of Science Position Statement (2021) on what all high school students should know about the NOS at graduation:

1. Scientific investigations use a variety of methods.
2. Scientific knowledge is based on empirical evidence.
3. Scientific knowledge is open to revision in light of new evidence.
4. Scientific models, laws, mechanisms, and theories explain natural phenomena.
5. Science is a way of knowing.
6. Scientific knowledge assumes an order and consistency in natural systems.
7. Science is a human endeavor.
8. Science addresses questions about the natural and material world.

PART IV

The Future of Real Science.

The future of real science lies in a more collaborative, transparent, interdisciplinary, and multisectoral approach, where the boundaries between scientific disciplines blur, and knowledge is shared openly (Varas et al., 2023). Advances in technology, computational science, and big data analytics will enable more precise and reproducible experiments, while open access to data and methodologies will foster global cooperation. Ethical considerations, coupled with better statistical practices, will ensure that scientific discoveries are both truthful and trustworthy. As real science evolves in the 21st century, it will increasingly address complex global challenges, leading to innovations that improve lives and deepen our understanding of the natural and material world around us.

CHAPTER 12

The Role of Technology in Advancing Real Science.

Technology plays a pivotal role in advancing real science by providing tools that enhance research capabilities, improve data accuracy, and accelerate discoveries (Oliveira et al., 2019). From sophisticated instruments that enable precise measurements to powerful computational models that simulate open and closed complex systems, technology allows scientists to explore realms previously inaccessible. It also facilitates global collaboration through digital platforms, enabling the seamless open sharing of data, ideas, and results across borders. As technology continues to evolve, it not only drives scientific innovation but also transforms how scientific research is conducted, making it more efficient, scalable, and inclusive.

12.1 The Collective Impact of Big Data, AI, and Machine Learning on Scientific Research.

The convergence of Big Data, Artificial Intelligence (AI), and Machine Learning (ML) is transforming scientific research in unprecedented ways (Ejjami, 2024).

Big Data allows researchers to analyze vast amounts of information from diverse sources, uncovering patterns and correlations that were previously impossible to detect. This abundance of data is fueling AI and ML algorithms, which can process and analyze these datasets at incredible speeds, offering real-time insights that drive scientific discoveries. AI and ML are particularly valuable in areas like genomics, phenomics, climate science, and drug discovery, where they can sift through complex datasets to identify potential breakthroughs or predict future trends.

Moreover, the integration of AI and ML with Big Data is enhancing the precision and efficiency of research (Althati et al., 2024). Machine learning models can learn from data and improve their accuracy over time, making predictions and generating hypotheses that guide further experimentation. In fields like personalized medicine (Udegbe et al., 2024) this means treatments can be tailored more accurately to individual patients based on their unique genetic information, personal biomarkers, and other health data. Similarly, in environmental science, AI-driven models can predict man-made climate change impacts with greater accuracy (Kumar et al., 2024), helping to inform public policy decisions.

The collective impact of Big Data, AI, and ML is also democratizing real science by making advanced analytical tools accessible to a broader range of researchers (Ritoré et al., 2024). Even those without expertise in data science can leverage these technologies to enhance their research, leading to more diverse and innovative contributions to the scientific community. As these technologies continue to evolve, they will further push the boundaries of what is possible in real science, enabling researchers to tackle complex global challenges with new levels of insight and efficiency.

12.2 The Promise and Perils of Citizen Science.

Citizen science holds great promise for open access to research and expanding the scope and scale of scientific inquiry (Serbe-Kamp et al., 2023; Jaeger et al., 2023). By engaging the public in data collection, analysis, and even hypothesis generation, citizen science projects can harness the collective power of thousands, or even millions, of volunteers. This approach is particularly valuable in fields like ecology, where large-scale data collection is often necessary. For example, citizen scientists can help monitor wildlife populations, track environmental changes, or even contribute to astronomical observations (Longdon et al., 2024). The involvement of non-experts can lead to the discovery of new phenomena, inspire public interest and trust in real science, and foster a greater sense of community engagement and empowerment around scientific endeavors.

However, citizen science also comes with potential perils that must be carefully managed (Quinn, 2020). The quality of data collected by non-experts can vary widely, leading to issues with validity and reliability. Without proper training and oversight, citizen scientists might misinterpret instructions, record incorrect data, or introduce implicit biases that can skew results. Additionally, there is a risk that the enthusiasm of volunteers may wane over time, leading to incomplete datasets or the need for continuous recruitment efforts.

Ethical considerations also arise in citizen science, particularly when involving vulnerable populations (Chesser et al., 2020) or sensitive environmental data (Fraisl et al., 2022). Ensuring that participants are fully informed about how their data will be used, and protecting their privacy, are critical to maintaining trust and truthfulness in these projects. Despite these challenges, with careful design, robust training, and ongoing support, citizen science can be a powerful tool for expanding the frontiers of research, while also engaging and empowering the public to contribute meaningfully to scientific knowledge.

12.3 How New Technologies are Reshaping the Landscape of Scientific Inquiry.

New technologies are fundamentally reshaping the landscape of scientific inquiry, transforming how research is conducted, shared, and applied (Kraus et al., 2021). Advances in areas like genomics, phenomics, nanotechnology, big data analysis, machine learning (ML), and artificial intelligence (AI) are pushing the boundaries of what is possible, enabling scientists to explore questions that were once out of reach. For instance, the advent of CRISPR-Cas9 technology has revolutionized genetic research (Bhatt et al., 2022), allowing for precise editing of DNA, which opens up new possibilities in fields ranging from medicine to agriculture. Similarly, nanotechnology is enabling the creation of materials and devices at the molecular and sub-molecular levels of organization, leading to breakthroughs in areas such as drug delivery, renewable energy, and electronics (Ahire et al., 2022).

The integration of artificial intelligence (AI) and machine learning (ML) into scientific research is another profound change. These technologies are enhancing the ability to analyze complex datasets, uncover patterns, and generate new hypotheses. In drug discovery, for example, AI can sift through vast amounts of data to identify potential compounds for further testing, significantly speeding up the pharmacologic research process (Nayarisseri et al., 2021; Cavasotto et al., 2021). AI and ML are also being used to create predictive models in fields like climate science (Zhong et al., 2021), epidemiology (Wiemken et al., 2020), and economics (Storm et al., 2020), allowing researchers to simulate scenarios and make more informed decisions.

Moreover, new technologies are democratizing scientific research by making advanced tools and resources more accessible to a wider range of researchers and even the public-at-large. Cloud computing, open-source software, and digital collaboration platforms (Lin et al., 2022) are breaking down barriers to entry, enabling scientists from around the world to collaborate in real-time, share data, and contribute to global research efforts. This increased accessibility is fostering creativity, innovation, and accelerating the pace of discovery, as diverse perspectives and expertise are brought together to tackle complex challenges.

However, these technological advancements also prompt critical questions about the future of scientific inquiry. The growing dependence on sophisticated tools and algorithms risks widening the gap between those with access to state-of-the-art technology and those without, potentially deepening inequalities within the scientific community. Moreover, the increasing complexity of research highlights the need for vigilance in areas such as cybersecurity, ethical and moral considerations, and rigorous validation of findings. As new technologies evolve, they will undoubtedly transform the scientific landscape, offering exciting opportunities for discovery while introducing new challenges that demand thoughtful navigation.

CHAPTER 13

Global Collaboration in Real Science.

Global collaboration in real science is becoming increasingly essential as the complexity of modern challenges demands diverse expertise and resources from around the world (InterAcademy Partnership, 2016). By bringing together scientists from different countries and disciplines, global collaboration fosters the exchange of ideas, accelerates innovation, and allows for the pooling of data and technologies. This cooperative and coordinated approach has been pivotal in addressing issues like man-made climate change, pandemics, natural disasters, and energy sustainability, where solutions require a coordinated global effort. As barriers and obstacles to communication and data sharing continue to fall, global collaboration will play an even more critical role in advancing scientific knowledge and driving impactful solutions for the world's most pressing problems (Holm et al., 2013).

13.1 The Collective Impact of International Collaboration in Solving Global Challenges.

International collaboration plays a pivotal role in addressing global challenges, acting as a catalyst for collective impact (Baker et al., 2024). In an increasingly and inextricably interconnected world, existential threats such as man-made climate change, pandemics, economic inequality and inequity, and institutional bias transcend national borders, requiring coordinated efforts from multiple nations. By pooling resources, knowledge, and expertise, countries can tackle these complex problems more effectively than any single nation could on its own (i.e., the whole is greater than the sum of the parts). International collaboration allows for the sharing of empirically-driven best practices, the harmonization of public policies, and the alignment of efforts toward common goals, fostering a sense of shared responsibility and global solidarity.

Moreover, international collaboration can bridge the gap between developed and developing nations, ensuring that solutions are inclusive and equitable (Samuel-Okon et al., 2024) Through partnerships (von Luepke et al., 2024), higher-income countries can support middle- and lower-income countries by providing financial aid, technological advancements, and capacity-building initiatives. This not only helps to address immediate crises but also strengthens and sustains the global community's ability to respond to future challenges. Collaborative efforts also lead to innovation (Audrelsch et al., 2024), as diverse perspectives and expertise converge to create more robust and sustainable solutions. Ultimately, the collective impact of international collaboration is a testament to the power of unity in addressing the world's most pressing issues.

13.2 Case studies: Climate Change, Pandemics, Global Commons, and Space Exploration.

13.2.1 Man-Made Climate Change.

Man-made climate change stands as a prominent example of the necessity and efficacy of international collaboration. The Paris Agreement, adopted in 2015 (Savaresi, 2016), exemplifies how global cooperation can lead to unified action against a shared threat. Nearly 200 countries committed to reducing their greenhouse gas emissions, setting a global framework for limiting global temperature rise to well below 2 degrees Celsius above pre-industrial levels. This collective effort has spurred national governments, private sectors, and civil societies to take concerted action. International collaborations like the Green Climate Fund further demonstrate how developed nations can support developing countries in their climate adaptation and mitigation efforts, ensuring that the global response to man-made climate change is inclusive and effective (Li et al., 2024).

13.2.2 Pandemics.

The SARS-CoV-2 (i.e., COVID-19) pandemic underscored the critical importance of international collaboration in managing global public health crises (Pennisi et al., 2024). The rapid spread of the virus highlighted the interconnectedness of the world and the need for a collaborative, cooperative, and coordinated response. Organizations, like the World Health Organization (WHO), played a central role in facilitating the sharing of information, resources, and best clinical and non-clinical practices across borders (Granata et al., 2024). The development and distribution of vaccines were also a testament to global cooperation, with initiatives like COVAX working to ensure that low- and middle-income countries had access to life-saving vaccines (Bostyn, 2024). The pandemic also sparked unprecedented global collaboration among scientists and researchers, rapidly advancing the development of critical diagnostics and treatments (Moneshwaran et al., 2024). Essential public health services were also expanded to an international level of organization across the globe for a more equitable, and real-time preparedness and response to an ever-changing structure of the SARS-CoV-2 virus (Patel et al., 2024).

13.2.3 Global Commons.

The concept of the Global Commons refers to natural resources and spaces that are shared by all nations, such as the high seas, the atmosphere, and the polar regions (Rockstrom et al., 2024). Managing and protecting these commons' areas requires international cooperation, as they are beyond the jurisdiction of any single nation. The

United Nations Convention on the Law of the Sea (UNCLOS) is a case study in how international agreements can govern the use and preservation of global commons (Armstrong, 2024). This treaty, adopted in 1982, established guidelines for the sustainable use of the oceans, balancing the needs of different countries with the imperative to protect marine environments. Similarly, the Antarctic Treaty System, which prohibits military activity and mineral mining on the continent while promoting scientific research, is another example of how nations can come together to manage shared resources responsibly (Chown et al., 2024)

13.2.4 Space Exploration.

Space exploration is another arena where international collaboration has been key to achieving significant milestones (Lambright, 2010). The International Space Station (ISS) is perhaps the most iconic example of this, serving as a symbol of peaceful cooperation between nations, including former adversaries (Kitmacher et al., 2010). The ISS is a joint project involving the United States, Russia, Europe, Japan, and Canada, where astronauts from different countries live and work together, conducting research that benefits humanity as a whole. This collaboration has not only advanced scientific knowledge but also fostered diplomatic relations between participating countries. As humanity looks toward future endeavors for the remainder of the 21st century, such as returning to the Moon and exploring Mars, international collaboration will be crucial in sharing the immense costs, risks, and benefits of space exploration (Reneau et al., 2021).

13.3 The Role of Real Science Diplomacy in Fostering Global Peace and Cooperation.

Real science diplomacy plays a crucial role in fostering global peace and cooperation by bridging the gap between scientific research and international public relations (Davis, 2014). It leverages scientific collaboration as a tool for building trust and truthfulness, enhancing culturally-competent dialogue, and addressing shared challenges among all nations. Through real science diplomacy, countries can engage in constructive partnerships, even in the face of political tensions, by focusing on common scientific goals and the pursuit of information and knowledge that transcends borders (Porsdam et al., 2024). This approach not only advances scientific understanding but also promotes mutual respect and understanding among nations, laying the groundwork for the collective impact of peaceful coexistence and collaborative problem-solving (e.g., social justice).

One of the critical aspects of real science diplomacy is its ability to facilitate dialogue on neutral grounds (Puaschunder, 2024). Scientific collaboration often takes place in a politically neutral environment, allowing countries to work together on issues of global importance, such as nature-based solutions (NbS) to climate change (Miralles-Wilhelm, 2021; Iseman et al., 2021), public health initiatives requiring essential functions and services (Falqui et al., 2024), and sustainable economic development (Puaschunder, 2024), without the influence of geopolitical rivalries. This can help to de-escalate conflicts and build channels of communication that may not be possible through traditional diplomatic means. Additionally, real science diplomacy fosters people-to-people connections (Pardeshi et al., 2024) through academic exchanges, joint research initiatives, public-private partnerships, and the sharing of scientific data, which can contribute to long-term sustainable relationships based on transparency, trust, truthfulness, and mutual benefit.

Furthermore, real science diplomacy is essential in addressing global challenges that require coordinated international efforts (Masters, 2024). Issues like pandemics, natural disasters, nuclear non-proliferation, and environmental degradation cannot be effectively tackled by individual countries acting alone. By encouraging scientific collaboration across borders, real science diplomacy enables the pooling of resources, expertise, and technology to develop innovative solutions to these complex problems. This collective approach not only enhances the effectiveness of global responses but also rein-forces the notion that real science can serve as a unifying force in a divided world (Theron, 2024).

In essence, real science diplomacy is a powerful tool for promoting global peace and cooperation. It enables countries to engage with one another on the basis of shared scientific goals, fostering trust, truthfulness, and understanding even in the face of political differences. By prioritizing collaboration over competition, real science diplomacy helps to build a more interconnected and peaceful world, where nations work in real time together to solve the most pressing global challenges of our time.

CHAPTER 14

The Path Forward: Ensuring the Integrity of Real Science.

Ensuring the integrity of real science is crucial for advancing knowledge and maintaining public trust (Bloome, 2024). To safeguard this integrity, it's essential to uphold rigorous standards of transparency, expert peer review, and reproducibility in scientific research. Addressing issues like data manipulation, conflicts of interest, and publication bias is vital for ensuring that scientific findings are reliable and credible. Moreover, fostering a culture of ethical conduct, transparency, and accountability among researchers and institutions can help prevent misconduct and promote the pursuit of truth (Dalkiran, 2024). By prioritizing these principles, we can protect the foundation of scientific inquiry and ensure that real science continues to contribute meaningfully to society and address global challenges.

14.1 Strategies for Promoting Transparency and Accountability in Scientific Research.

Promoting transparency and accountability in scientific research is essential for maintaining the credibility and reliability of scientific findings. One effective strategy is to encourage open access to research data and methodologies (Gao et al., 2023). By making raw data, analysis scripts, and experimental protocols publicly available, researchers can allow others to verify, replicate, and build upon their work. This openness not only enhances the reproducibility of results but also fosters a culture of trust, truthfulness, collaboration, and accountability within the scientific community.

Another critical strategy is the implementation of rigorous expert peer review processes (Waltman et al., 2023). Ensuring that research undergoes thorough evaluation by independent subject matter experts before publication helps to identify potential errors or biases in the study. Additionally, promoting pre-registration of research protocols and hypotheses can prevent selective reporting and p-hacking (Brodeur et al., 2023), where only favorable results are published. This practice establishes clear expectations and provides a framework for assessing the validity and reliability of the research findings.

Institutional policies and funding bodies also play a key role in promoting transparency and accountability (Adeusi et al., 2024) Research institutions should enforce ethical guidelines and conduct regular audits to detect and address issues of misconduct. Funding agencies can support these efforts by requiring transparency and accountability measures as conditions for grants. Furthermore, providing training and resources on research ethics and empirically-derived best practices helps researchers adhere to the highest standards of integrity and responsibility throughout their careers.

Lastly, fostering a culture of openness and ethical behavior within the scientific community is truly foundational (Roy et al., 2024). Encouraging dialogue about the importance of research integrity, celebrating positive examples of transparency, and holding individuals and institutions accountable for breaches of ethical conduct can collectively strengthen the commitment to high-quality real science world-wide. By integrating these strategies, the scientific community can enhance the validity, reliability, and reproducibility of research outcomes and uphold the trust of the public.

14.2 The Importance of Global Interdisciplinary Collaboration and Cross-sector Partnerships.

Global interdisciplinary collaboration and cross-sector partnerships are increasingly vital in addressing the complex and multifaceted challenges of today's world (Sultan, 2024; Jalal, 2024). Many of the most pressing issues, such as man-made climate change, public health crises, and technological innovation, span multiple disciplines and sectors. Addressing these challenges effectively requires expertise from a range of fields—scientists, engineers, economists, policymakers, and social scientists–who must work together collaboratively to develop comprehensive, sustainable global solutions. Interdisciplinary and multisectoral collaboration allows for the integration of diverse perspectives, leading to more creative, innovative, integrated, and holistic approaches.

Cross-sector partnerships further enhance the impact of interdisciplinary efforts by bringing together various stakeholders, including government agencies, private industry, non-governmental organizations, and academic institutions (Dzhengiz et al., 2024). These partnerships leverage the unique resources and strengths of each sector, fostering a collaborative environment that can tackle complex problems more efficiently. For example, in addressing global health issues, partnerships between public health organizations, pharmaceutical companies, and local communities can accelerate the development and distribution of vaccines (Nunes et al., 2024), as well as improve health outcomes in underserved and marginalized areas.

Additionally, global interdisciplinary and multisector collaborations can help to bridge gaps between education, research and practice (Sultan, 2024). By involving stakeholders from different sectors early in the research process, solutions can be designed with practical implementation in mind, ensuring that they are feasible and sustainable. These partnerships also facilitate the sharing of knowledge and resources across borders, enabling more effective responses to global challenges. Overall, the importance of global interdisciplinary collaboration and cross-sector partnerships lies in their ability to create synergies that drive progress and address complex issues with greater efficacy and innovation.

14.3 Ensuring that Real Science Remains a Tool for Truth, Trustworthiness, and Progress in the 21st century.

Maintaining real science as a tool for truth, trustworthiness, and progress in the 21st century requires a comprehensive approach grounded in the fundamental principles of scientific inquiry (Resnik et al., 2023). A key element of this effort is a steadfast commitment to research integrity, which emphasizes transparency, accountability, reproducibility, and rigorous expert peer review. By upholding these standards, the scientific community can minimize errors, biases, and misconduct, ensuring the reliability and validity of research outcomes. Additionally, promoting the interdisciplinary and multi-sectoral sharing of research data and methodologies enhances transparency and accountability, enabling independent verification and strengthening public trust in the scientific process.

Another crucial component is the effective communication of scientific knowledge to the public (Bayes et al., 2023). In an age where information, misinformation, and disinformation spreads rapidly in real time (e.g., 24-7 news cycle), scientists must actively engage with diverse audiences, presenting findings clearly and making their significance easy to grasp. By educating the public on the scientific method and the inherent limitations of scientific knowledge, society becomes better equipped to critically assess information and make informed, evidence-based decisions, with or without expert guidance. This openness not only nurtures public trust but also ensures the responsible use of scientific advancements.

Furthermore, real science must continuously evolve to meet the complex challenges of the 21st century by embracing interdisciplinary and multisectoral approaches (Cooke et al., 2024). Addressing global existential threats—such as human-driven climate change, natural disasters, and pandemics—demands expertise from multiple fields and the integration of diverse perspectives. By fostering cross-sector partnerships and global collaborations, real science can develop more effective solutions, drive innovation, and promote sustainable progress that benefits both society and the planet.

Ultimately, preserving real science as a tool for truth, trust, and progress demands a collective effort from the scientific community, policymakers, and the public. Upholding rigorous research standards, communicating findings clearly, and fostering collaboration are crucial to ensuring that real science remains a vital force in addressing the critical challenges of our time and contributing meaningfully to society.

CHAPTER 15

Conclusion:
Reaffirming the Pursuit
of Real Science.

Reaffirming the pursuit of real science involves a steadfast commitment to the principles of empirically-driven and derived evidence, rigorous methodology, and objective inquiry. It means prioritizing evidence-based research over sensationalism and ensuring that scientific practices are guided by integrity and transparency. This pursuit demands a focus on replicability, expert peer review, and open dialogue, which collectively help to safeguard against institutional biases, disinformation, and misinformation. By upholding these foundational standards, we can ensure that real science remains a reliable tool for advancing information and knowledge, solving real-world problems, and fostering progress in an increasingly complex, complicated, and rapidly changing world.

15.1 Summarizing Key Takeaways from the Book

In the Pursuit of Real Science in a Modern-Day Society, the book underscores several key principles vital to preserving the integrity and impact of scientific inquiry today. A central theme is the unwavering commitment to scientific rigor and objectivity. The book also calls for strict adherence to the core principles of validity, reliability, and reproducibility, all crucial to ensuring the credibility and trustworthiness of scientific findings. It also emphasizes the importance of rigorous expert peer review and the democratization of data to promote transparency and accountability and curb the spread of misinformation and disinformation.

Another significant takeaway is the role of effective science communication and capacity-building in bridging the gap between researchers and the public-at-large. The book underscores the necessity for scientists to engage with diverse audiences, convey complex findings in simpler terms, and combat the deliberate release of misinformation for political and financial gain. By enhancing public understanding of the scientific method and the limitations of disciplined research, real science can be better appreciated and applied in decision-making processes.

In the Pursuit of Real Science in a Modern-Day Society addresses the importance of interdisciplinary and multisectoral approaches in tackling complex global challenges. It stresses that many modern issues, such as man-made climate change and public health crises, require the integration of valid and reliable information and knowledge from various fields and sectors. Building cross-disciplinary and cross-sector partnerships can lead to more innovative and comprehensive solutions in serving the public-at-large.

In the Pursuit of Real Science in a Modern-Day Society serves as an unwavering call to action for upholding the highest global standards in scientific research while fostering a culture of transparency, rigorous inquiry, and ethical responsibility. In an era marked by rapid technological advancements, complex global crises, and the proliferation of misinformation, this work emphasizes the critical need for improved communication, interdisciplinary and multisectoral collaboration, and a shared commitment to truth and progress.

By reinforcing these principles, the scientific community can remain a reliable, forward-thinking force in addressing the most pressing existential threats to all living things—ranging from man-made climate change and global pandemics to artificial intelligence ethics and biotechnological integrity. The pursuit of real science demands not only technical excellence but also public engagement and policy advocacy to ensure that scientific advancements serve humanity's best interests. Now, more than ever, a renewed dedication to intellectual integrity, open discourse, and empirically-derived evidence-based decision-making is essential to shaping a sustainable and resilient future.

15.2 The Enduring Importance of Skepticism, Critical Thinking, Logical Reasoning, and Ethical Considerations.

The enduring importance of skepticism, critical thinking, logical reasoning, and ethical considerations cannot be overstated in the realm of real science and beyond. Skepticism is fundamental to the scientific process, driving researchers to question assumptions, challenge established norms and beliefs, and seek evidence before accepting claims. This healthy doubt ensures that scientific knowledge is not taken at face value but is rigorously tested and validated, thus enhancing the reliability of findings.

Critical thinking further amplifies this approach by encouraging a systematic evaluation of information. It involves analyzing arguments, identifying biases, and assessing the validity of evidence, which helps to differentiate between well-supported conclusions and those based on flawed reasoning. In a world where misinformation and pseudoscience can spread rapidly, critical thinking is essential for making informed decisions and fostering a deeper understanding of complex issues.

Logical reasoning complements these practices by providing a framework for constructing coherent and consistent arguments during discourse. It helps to ensure that conclusions are drawn based on sound principles and empirically-driven evidence, preventing logical fallacies that can undermine the credibility of scientific and intellectual discussion. Logical reasoning is crucial for developing robust theories and solving problems effectively.

Ethical considerations are equally foundational, as they guide the responsible conduct of research and the application of scientific information and knowledge. Upholding ethical standards helps to prevent misconduct, protect the welfare of participants, and ensure that research outcomes are used for the greater good. By integrating ethical principles with skepticism, critical thinking, and logical reasoning, we can uphold the integrity of scientific inquiry and contribute positively to society.

15.3 A Collective Call to Action: Advancing Real Science for a Sustainable Future.

A unified call to action for scientists, educators, policymakers, and the public-at-large is essential for advancing the pursuit of real science and ensuring its transformational impact on society. In an era defined by rapid technological advancements, emerging global threats, and widespread misinformation, a steadfast commitment to scientific integrity and public engagement is more critical than ever.

15.3.1 Scientists: Leading with Integrity and Transparency.

Scientists must set the standard for research integrity, leading by example and upholding the highest ethical and methodological principles. A strong commitment to transparency, reproducibility, and rigorous expert peer review is necessary to maintain public trust and ensure that scientific knowledge progresses in an ethical and credible manner. Scientists should actively engage in open data-sharing initiatives, interdisciplinary and multisectoral collaborations, and science communication efforts to enhance public understanding and counter misinformation. Furthermore, fostering a culture of collegiality and truthfulness within the scientific community is essential to encouraging intellectual exchange, constructive debate, and the collective pursuit of knowledge.

15.3.2 Educators: Cultivating Scientific Literacy and Critical Thinking.

Educators play a crucial role in shaping a scientifically literate society by instilling the values of critical thinking, logical reasoning, and empirical inquiry in their students. The integration of real-world scientific issues into curricula, combined with inquiry-based learning approaches, fosters curiosity and deepens students' understanding of complex scientific concepts. By emphasizing the importance of the scientific method and equipping students with the skills to critically evaluate information, educators empower the next generation of students to become informed citizens and future innovators. Moreover, embedding discussions on capacity-building and sustainability within scientific education ensures that students recognize their role in shaping a more resilient and responsible global community.

15.3.3 Policymakers: Enabling Science-Driven Decision-Making.

Policymakers hold the responsibility of creating an environment that supports robust scientific research, ensuring its applications benefit society at large. This includes securing adequate and sustainable funding for scientific endeavors, establishing policies that promote transparency and ethical research practices, and integrating scientific evidence into decision-making at all levels of governance. By fostering partnerships between government agencies, academic institutions, and industry leaders, policymakers can help bridge the gap between research and its real-world applications. Additionally, prioritizing science communication efforts and public engagement initiatives can enhance trust in scientific institutions and encourage evidence-based policymaking.

15.3.4 Public-at-Large: Engaging with Science for a Better Future.

The public-at-large also has a pivotal role to play in supporting and advocating for real science. Staying informed about scientific developments, critically assessing information sources, and participating in discussions about science-related issues are fundamental to fostering a more engaged and knowledgeable society. Individuals can contribute to scientific progress by supporting public science initiatives, advocating for policies that prioritize research, and promoting real science education in their communities. A society that values and understands real science is better equipped to address global challenges, from man-made climate change to public health crises, ensuring a future where scientific advancements serve humanity's best interests.

15.3.5 The 4 C's: Collaboration, Cooperation, Coordination, and Communication.

Advancing real science and harnessing its full potential requires a collective effort from all stakeholders. By embracing the 4 C's—Collaboration, Cooperation, Coordination, and Communication—scientists, educators, policymakers, and the public-at-large can work together to build a robust scientific ecosystem that fosters creativity, innovation, integrity, and accessibility. Through these concerted efforts, real science will continue to be a driving force in solving the most pressing STEM (science, technology, engineering, and mathematics) issues of today and shaping a more informed, resilient, and forward-thinking society for generations to come.

REFERENCES.

Abd-El-Khalick, F. (2012) Nature of science in science education: Toward a coherent framework for synergistic research and development. *Second international Handbook of Science Education* 1041-1060.

Adelekan, O. A., Ilugbusi, B. S., Adisa, O., et al. (2024) Energy transition policies: a global review of shifts towards renewable sources. *Engineering Science & Technology Journal* 5(2): 272-287.

Adenyi, A. O., Okolo, C. A., Olorunsogo, T., et al. (2024) Leveraging big data and analytics for enhanced public health decision-making: A global review. *GSC Advanced Research and Reviews* 18(2): 450-456.

Adeusi, K. B., Jejeniwa, T. O., Jejeniwa, T. O. (2024) Advancing financial transparency and ethical governance: innovative cost management and accountability in higher education and industry. *International Journal of Management & Entrepreneurship Research* 6(5): 1533-1546.

Adler, R. H. (2022) Trustworthiness in qualitative research. *Journal of Human Lactation* 38(4): 598-602.

Ahire, S. A., Bachhav, A. A., Pawar, T. B., et al. (2022) The Augmentation of nanotechnology era: A concise review on fundamental concepts of nanotechnology and applications in material science and technology. *Results in Chemistry* 4: 100633.

Alai, M. (2017) The debates on scientific realism today: knowledge and objectivity in science. *In* Varieties of Scientific Realism: Objectivity and Truth in Science.

Alexandrova, A. (2017) *A philosophy for the science of well-being.* Oxford University Press: Oxford, England.

Al Hamad, N. M., Adewusi, O. E., Unachukwu, C. C., et al. (2024) Integrating human resources principles in STEM education: A review. *World Journal of Advanced Research and Reviews* 21(1): 1174-1183.

Alismail, H. A., McGuire, P. (2015) 21st century standards and curriculum: Current research and practice. *Journal of Education and Practice* 6(6): 150-154.

Allchin, D. (2011) Evaluating knowledge of the nature of (whole) science. *Science Education* 95(3): 518-542.

Allison, L. A. (2021) *Fundamental molecular biology.* John Wiley & Sons.

Allum, N. (2011) What makes some people think astrology is scientific? *Science Communication* 33(3): 341-366.

Almasri, F. (2024) Exploring the impact of artificial intelligence in teaching and learning of science: A systematic review of empirical research. *Research in Science Education* 54(5): 977-997.

Althati, C., Tomar, M., Shanmugam, L. (2024) Enhancing Data Integration and Management: The Role of AI and Machine Learning in Modern Data Platforms. *Journal of Artificial Intelligence General Science* 2(1): 220-232.

Alvesson, M., Sandberg, J. (2024) The art of phenomena construction: A framework for coming up with research phenomena beyond the usual suspects. *Journal of Management Studies* 61(5): 1737-1765.

Amfo, N. A. A., Awandare, G. A. (2024) Equitable Collaborations: Key to Sustainable Futures. *In* The new road to success: Contributions of universities towards more resilient societies.

Amir-Azodi, A., Setayesh, M., Bazyar, M., et al. (2024) Causes and consequences of quack medicine in health care: a scoping review of global experience. *BMC Health Services Research* 24(1): 64.

Anderson, S. T., Lin, K. K. (2024). Scientific method. *In* Translational Orthopedics. Academic Press.

Andrew, A. (2024) Understanding the "Infodemic" Threat: A Case Study of the COVID-19 Pandemic. *Journal of the Korean Academy of Family Medicine.*

Anjum, R. L., Rocca, E. (2024). Scientific Methods and Causal Evidencing. Bias about Causality. *In* Philosophy of Science. Cham: Springer International Publishing.

Anjum, R. L., Rocca, E. (2024b) A Dispositional Account of Causation. *Alternative Approaches to Causation: Beyond Difference-Making and Mechanism.*

Anjum, R. L., Rocca, E. (2024c). Is Science Defined by Its Community? *In* Philosophy of Science. Cham: Springer International Publishing.

Antony, J. (2023) *Design of experiments for engineers and scientists.* Elsevier.

Antunes, B. A., Hill, D. R. (2024) Reproducibility, Replicability, and Repeatability: A survey of reproducible research with a focus on high performance computing.

Arabatzis, T., Schickore, J. (2012) Ways of integrating history and philosophy of science. *Perspectives on Science* 20(4): 395-408.

Armstrong, C. (2024) The United Nations Convention on the Law of the Sea, global justice and the environment. *Global Constitutionalism* 13(1): 16-20.

Ashcraft, L. E., Cabrera, K. I., Lane-Fall, M. B., et al. (2024) Leveraging implementation science to advance environmental justice research and achieve health equity through neighborhood and policy interventions. *Annual Review of Public Health* 45.

Atkins, P. W., Ratcliffe, R. G., de Paula, J., et al. (2023) *Physical chemistry for the life sciences.* Oxford University Press.

Audretsch, D. B., Belitski, M. (2024). Knowledge collaboration, firm productivity and innovation: A critical assessment. *Journal of Business Research* 172: 114412.

Avery, J. S. (2021) *Information theory and evolution.* World Scientific.

Bailer-Jones, D. M. (2009) *Scientific models in philosophy of science.* University of Pittsburgh Press.

Baker, D. P., Powell, J. J. (2024) *Global Mega-Science: Universities, Research Collaborations, and Knowledge Production.* Stanford University Press.

Baum, C. M., Bass, J. D., Christiansen, C. H. (2024). Theory, models, frameworks, and classifications. In *Occupational Therapy*. Routledge.

Bayes, R., Bolsen, T., Druckman, J. N. (2023) A research agenda for climate change communication and public opinion: The role of scientific consensus messaging and beyond. *Environmental Communication* 17(1): 16-34.

Becker, A., Lukka, K. (2023). Instrumentalism and the publish-or-perish regime. *Critical Perspectives on Accounting* 94: 102436.

Bendiscioli, S., Firpo, T., Bravo-Biosca, A., et al. (2023) The experimental research funder's handbook.

Bennett, N. J., Whitty, T. S., Finkbeiner, E., et al. (2018). Environmental stewardship: A conceptual review and analytical framework. *Environmental Management* 61: 597-614.

Betts, K. R., Aikin, K. J., Miles, S., et al. (2024) Disease Awareness and Prescription Drug Communications on Television: Evidence for Conflation and Misleading Product Impressions. *Health Communication*.

Beuchert, T., Cayless, A., Darbellay, F., et al. (2024) Science for society. *In* EPS Grand Challenges: Physics for Society in the Horizon 2050. IOP Publishing.

Bhatt, P., Singh, S., Alfuraiji, N., et al. (2022) CRISPR CAS9: A new technology to modify genome-A review. *Open J. Syst. Demonstr J* 8(4): 208-215.

Biermann, F., Hickmann, T., Sénit, C. A., et al. (2022) Scientific evidence on the political impact of the Sustainable Development Goals. *Nature Sustainability* 5(9): 795-800.

Blomme, L. (2024). Objectivity and Promoting Public Trust in Value-Laden Science.

Bloor, D. (1976) *Knowledge and Social Imagery*. University of Chicago Press: Chicago, IL.

Bonyadi, F., Khaniki, H., Zardar, Z. (2024) Mediated Trust in Science: Revisiting the Role of Science Communication as Intermediary

Communication in Trust in Science. *Interdisciplinary Studies in Media and Culture.* 14(1).

Borsa, A., Bejarano, G., Ellen, M., et al. (2023) Evaluating trends in private equity ownership and impacts on health outcomes, costs, and quality: systematic review. BMJ.

Bostyn, S. J. (2024) Access to drugs, patents, and pandemic crisis: A tale of (non-) inclusivity. *In* Research Handbook on Intellectual Property Rights and Inclusivity. Edward Elgar Publishing.

Bowen, W. M., Gleeson, R. E., Bowen, W. M., et al. (2019) The Industrial Revolution and Its Effects. *The Evolution of Human Settlements: From Pleistocene Origins to Anthropocene Prospects.*

Bowser, G., Ho, S. S., Ziebell, A., et al. (2024) Networking and collaborating: the role of partnerships across sectors to achieve educational goals in sustainability. *Sustainable Earth Reviews* 7(1): 17.

Boylan, J. E. (2016). Reproducibility. *IMA Journal of Management Mathematics, 27*(2): 107-108.

Briant, E. L. (2024) Researching and Conceptualizing the Influence of Industry. *Routledge Handbook of the Influence Industry.*

Brodeur, A., Cook, N. M., Hartley, J. S., et al. (2024). Do Preregistration and Preanalysis Plans Reduce p-Hacking and Publication Bias? Evidence from 15,992 Test Statistics and Suggestions for Improvement. *Journal of Political Economy Microeconomics* 2(3): 527-561.

Bromme, R., Hendriks, F. (2024). Trust in science: considering whom to trust for knowing what is true. *In* A Research Agenda for Trust. Edward Elgar Publishing.

Brookfield, S. D. (2011) *Teaching for critical thinking: Tools and techniques to help students question their assumptions.* John Wiley & Sons.

Brinsfield, T. N., Pinson, N. R., Levine, A. D. (2024) The evolution and ongoing challenge of unproven cell-based interventions. *Stem Cells Translational Medicine.*

Britt, M. A., Richter, T., Rouet, J. F. (2014) Scientific literacy: The role of goal-directed reading and evaluation in understanding scientific information. *Educational Psychologist* 49(2): 104-122.

Brock, S., Mares, E. (2014) *Realism and Anti-realism*. Routledge.

Bromme, R., Hendriks, F. (2024) Trust in science: considering whom to trust for knowing what is true. *In* A Research Agenda for Trust. Edward Elgar Publishing.

Brumback, A. C., Ngiam, W. X., Lapato, D. M., et al. (2024). Catalyzing communities of research rigour champions. *Brain Communications* 6(3): fcae120.

Buede, D. M., Miller, W. D. (2024). *The Engineering Design of Systems: Models and Methods*. John Wiley & Sons.

Buttar, A. M., Arshad, M. N., Nayyar, A. (2024) Exploring the Intersection of Biology and Computing: Road Ahead to Bioinformatics. *Artificial Intelligence and Machine Learning in Drug Design and Development*.

Butterworth, J., Smerdon, D., Baumeister, R., et al. (2024) Cooperation in the time of COVID. *Perspectives on Psychological Science* 19(4): 640-651.

Buzbas, E. O., Devezer, B., & Baumgaertner, B. (2023) The logical structure of experiments lays the foundation for a theory of reproducibility. *Royal Society Open Science* 10(3): 221042.

Cantner, U., Kalthaus, M., Yarullina, I. (2024) Outcomes of science-industry collaboration: Factors and interdependencies. *The Journal of Technology Transfer* 49(2): 542-580.

Cao, Y., Chen, R. C., Katz, A. J. (2024) Why is a small sample size not enough? *The Oncologist*.

Capalbo, A., de Wert, G., Mertes, H., et al. (2024) Screening embryos for polygenic disease risk: a review of epidemiological, clinical, and ethical considerations. *Human Reproduction Update* dmae012.

Caron, R. M., Hewitt, A. M. (2024) Population health: An integrating approach for aligning public health and healthcare systems. *Journal of Health Administration Education* 40(2): 231-246.

Casula, M., Rangarajan, N., Shields, P. (2021) The potential of working hypotheses for deductive exploratory research. *Quality & Quantity* 55(5): 1703-1725.

Caudill, D. S. (2023) *Expertise in crisis: The ideological contours of public scientific controversies.* Policy Press.

Cavasotto, C. N., Di Filippo, J. I. (2021) Artificial intelligence in the early stages of drug discovery. *Archives of Biochemistry and Biophysics* 688: 108730.

Chakravartty, A. (2023) Scientific knowledge vs. knowledge of science: Public understanding and science in society. *Science & Education* 32(6): 1795-1812.

Chakravorti, T., Koneru, S. D., Rajtmajer, S. (2024) Reproducibility, Replicability, and Transparency in Research: What 430 Professors Think in Universities across the USA and India.

Chalmers, A. (2013) *What is this thing called science?* McGraw-Hill Education: UK.

Chang, M. (2014). *Principles of scientific methods.* CRC Press.

Chen, E. K. (2025) Laws of Physics. *Elements in the Philosophy of Physics.*

Chesser, S., Porter, M. M., Tuckett, A. G. (2020) Cultivating citizen science for all: Ethical considerations for research projects involving diverse and marginalized populations. *International Journal of Social Research Methodology* 23(5): 497-508.

Chou, C., Lee, I. J., Fudano, J. (2024). The present situation of and challenges in research ethics and integrity promotion: Experiences in East Asia. *Accountability in Research* 31(6): 576-599.

Choudhary, E., Narayanan, S., Khan, F. (2024) Legal and Regulatory Landscape. *In* AI Healthcare Applications and Security, Ethical, and Legal Considerations. IGI Global.

Chown, S. L., Bastmeijer, K., Brooks, et al. (2024) Science advice for international governance–An evidence-based perspective on the role of SCAR in the Antarctic Treaty System. *Marine Policy* 163: 106143.

Christie, C. A., Alkin, M. C. (2023) An evaluation theory tree. *In* Evaluation Roots: Theory Influencing Practice.

Chu, Z., Li, S. (2023) Causal effect estimation: Recent progress, challenges, and opportunities. *Machine Learning for Causal Inference.*

Clark, B. R. (2023) *Places of inquiry: Research and advanced education in modern universities.* Univ of California Press.

Coccia, M. (2018) An introduction to the methods of inquiry in social sciences. *Journal of Social and Administrative Sciences* 5(2): 116-126.

Cochrane, K. L., Butterworth, D. S., Hilborn, R., et al. (2024) Errors and bias in marine conservation and fisheries literature: Their impact on policies and perceptions. *Marine Policy* 168: 106329.

Cole, S. (1992) *Making science: Between nature and society.* Harvard University Press.

Cole, N. L., Kormann, E., Klebel, T., et al. (2024) The societal impact of Open Science: a scoping review. *Royal Society Open Science* 11(6): 240286.

Collins, H., Evans, R. (2017) *Why democracies need science.* John Wiley & Sons.

Collins, S. (2023) *Neuroscience for learning and development: How to apply neuroscience and psychology for improved learning and training.* Kogan Page Publishers.

Cooke, S. J., Arlinghaus, R. (2024) Learning, thinking, sharing, and working across boundaries in fisheries science. *ICES Journal of Marine Science* 81(4): 665-675.

Cornelissen, J. P. (2023) The problem with propositions: Theoretical triangulation to better explain phenomena in management research. *Academy of Management Review.*

Cozzo, C. (2023) Fallibility and fruitfulness of deductions. *Erkenntnis* 88(7): 2997-3013.

Cozzoli, D. (2024) Kuhn, Popper, the Military-Industrial Complex and the Techno-scientific Revolution. *In* Rethinking Thomas Kuhn's Legacy. Cham: Springer International Publishing.

Crawford, L. The project sponsor. *In* The Handbook of Project Management. Routledge.

Crossman, J. (2024) The Relationship Between Religion, Superstition and Spirituality. In Superstition, Management and Organisations: Irrationality, Randomness, and Chaos in Decision Making. Cham: Springer Nature Switzerland.

Cullinan, M. E., Zimdars, M., Na, K. (2024) Their Truth is Out There: Scientific (Dis) trust and Alternative Epistemology in Online Health Groups. *Social media+ Society* 10(3).

Dalkıran, A. Guidelines for Scientific Research and Publication Ethics.

Danielson, R. W., Jacobson, N. G., Patall, E. A., et al. (2024) The effectiveness of refutation text in confronting scientific misconceptions: A meta-analysis. *Educational Psychologist.*

Darwin, C. (1859) *On the Origin of Species by Means of Natural Selection, or the Preservation of Favoured Races in the Struggle for Life.* John Murray, London.

Daubs, M. S. (2024) Wellness communities and vaccine hesitancy. *Media International Australia.*

Davidson, E. H. (2015) Genomics," Discovery Science," Systems Biology, and Causal Explanation: What Really Works? *Perspectives in Biology and Medicine,* 58(2): 165-181.

Davis, L. S. (2014) *Science diplomacy: New day or false dawn?* World Scientific.

Dehalwar, K., Sharma, S. N. (2024) Exploring the Distinctions between Quantitative and Qualitative Research Methods. *Think India Journal* 27(1): 7-15.

Deng, J. M., Ahmed, S. E., Awoonor-Williams, E., et al. (2024) Prioritizing mentorship as scientific leaders. *ACS Central Science* 10(2): 209-213.

Desmond, H., Ariew, A., Huneman, P., et al. (2024) The Varieties of Darwinism: Explanation, Logic, and Worldview. *The Quarterly Review of Biology* 99(2): 77-98.

Develaki, M. (2024) Uncertainty, Risk, and Decision-Making: Concepts, Guidelines, and Educational Implications. *Science & Education.*

DeVerna, M. R., Guess, A. M., Berinsky, A. J., et al. (2024) Rumors in retweet: Ideological asymmetry in the failure to correct misinformation. *Personality and Social Psychology Bulletin* 50(1): 3-17.

Di Ciccio, C., Marrella, A., Russo, A. (2015) Knowledge-intensive processes: characteristics, requirements and analysis of contemporary approaches. *Journal on Data Semantics* 4: 29-57.

Dzhengiz, T., Patala, S. (2024) The role of cross-sector partnerships in the dynamics between places and innovation ecosystems. *R&D Management* 54(2): 370-397.

Donley, D. W. (2024) Teaching the Nature of Science Improves Scientific Literacy Among Students Not Majoring in STEM. *J Undergrad Neurosci Educ* 22(2): A147-A152.

Dozier, S. J., MacPherson, A., Morell, L., et al. (2023) A Learning Progression for Understanding Interdependent Relationships in Ecosystems. *Sustainability* 15(19): 14212.

Dubin, J. A., Hameed, D., Baksh, N., et al. (2024) Impact of Reporting Bias, Conflict of Interest, and Funding Sources on Quality of Orthopaedic Research. *The Journal of Arthroplasty* 39(5): 1348-1352.

Eastwell, P. (2010) The scientific method: Critical yet misunderstood. *Science Education Review* 9(1): 8- 12.

Edwards-Price, S. (2021) *The long and winding road to reflexology: A post-structural narrative inquiry.*

Ejjami, R. (2024) Revolutionizing Research Methodologies: The Emergence of Research 5.0 through AI, Automation, and Blockchain. *IJFMR* 6(4): 1-33.

Elabbar, A. (2024) Expertise, moral subversion, and climate deregulation. *Synthese* 203(5): 1-28.

El-Sherbini, T. (2024) The Evolution of Physicists' Perception of the Universe. *Egyptian Journal of Physics* 52(1): 1-13.

Enquist, B. J., Kempes, C. P., West, G. B. (2024) Developing a predictive science of the biosphere requires the integration of scientific cultures. *Proceedings of the National Academy of Sciences* 121(19): e2209196121.

Erduran, S. (2023) Social and institutional dimensions of science: The forgotten components of the science curriculum? *Science* 381(6659): eadk1509.

Fahnestock, J. (2019) Rhetorical citizenship and the science of science communication. In Rhetoricians on Argumentation. Cham: Springer Nature Switzerland.

Falk, C. F., Muthukrishna, M. (2023) Parsimony in model selection: Tools for assessing fit propensity. *Psychological Methods* 28(1): 123.

Falqui, L., Li, F., Xue, Y. (2024) Global health diplomacy in humanitarian action. *Conflict and Health* 18(1): 46.

Feng, Q., Li, Q., Zhou, H., et al. (2024) CRISPR technology in human diseases. *MedComm* 5(8): e672.

Ferngren, G. B. (2022) Science and religion. *In* The Routledge History of American Science. Routledge.

Fife, S. T., Gossner, J. D. (2024) Deductive qualitative analysis: Evaluating, expanding, and refining theory. *International Journal of Qualitative Methods* 23: 16094069241244856.

Fišar, M., Greiner, B., Huber, C., et al. (2024) Reproducibility in Management Science. *Management Science* 70(3): 1343-1356.

Fleury, M. J. (2024) The Zilch-Zitter Electron.

Fliessbach, T. (2024) *Mechanics for Physicists: An Introduction, Including Special Relativity*. World Scientific.

Forscher, P. S., Wagenmakers, E. J., Coles, N. A., et al. (2023) The benefits, barriers, and risks of big-team science. *Perspectives on Psychological Science* 18(3): 607-623.

Fortunato, S., Bergstrom, C. T., Börner, K., et al. (2018) Science of science. *Science* 359(6379): eaao0185.

Fraisl, D., Hager, G., Bedessem, B., et al. (2022) Citizen science in environmental and ecological sciences. *Nature Reviews Methods Primers* 2(1): 64.

Furnari, S., Crilly, D., Misangyi, V. F., et al. (2021). Capturing causal complexity: Heuristics for configurational theorizing. *Academy of Management Review* 46(4): 778-799.

Gaida, M. E. (2022) Heliocentrism. *In* Encyclopedia of Renaissance Philosophy. Springer International Publishing.

Galdames, I. S., de Toro, X., Acevedo, D. (2024) Bridging Science and Society: The Role of University Science Communication Centers. *European Journal of Education and Psychology* 17(1): 1-23.

Gao, Y., Janssen, M., Zhang, C. (2023) Understanding the evolution of open government data research: towards open data sustainability and smartness. *International Review of Administrative Sciences* 89(1): 59-75.

Gatto, L., Aebersold, R., Cox, J., et al. (2023) Initial recommendations for performing, benchmarking and reporting single-cell proteomics experiments. *Nature Methods* 20(3): 375-386.

Gauchat, G. W. (2023) The legitimacy of science. *Annual Review of Sociology* 49(1): 263-279.

Gauthier, J., Vincent, A. T., Charette, S. J., et al. (2019) A brief history of bioinformatics. *Briefings in Bioinformatics* 20(6): 1981-1996.

Ghobakhloo, M., Iranmanesh, M., Tseng, M. L., et al. (2023). Behind the definition of Industry 5.0: A systematic review of technologies, principles, components, and values. *Journal of Industrial and Production Engineering* 40(6): 432-447.

Glicksman, R. L., Buzbee, W. W., Mandelker, D. R., et al. (2023) *Environmental Protection: Law and Policy [Connected EBook with Study Center]*. Aspen Publishing.

Godfrey, H., Erduran, S. (2023) Argumentation and intellectual humility: a theoretical synthesis and an empirical study about students' warrants. *Research in Science & Technological Education* 41(4): 1350-1371.

Goel, H., Raheja, D., Nadar, S. K. (2024) Evidence-based medicine or statistically manipulated medicine? Are we slaves to the P-value? *Postgraduate Medical Journal*.

Goldsteen, R. L., Goldsteen, K., Dwelle, T. (2024) *Introduction to Public Health: Promises and Practices.* Springer Publishing Company.

Gomes, C. M., Marchini, G., Bessa, et al. (2024). The landscape of biomedical research funding in Brazil: a current overview. *International Braz J Urol* 50(2): 209-222.

Gooding, D. C. (2012) *Experiment and the making of meaning: Human agency in scientific observation and experiment* (Vol. 5). Springer Science & Business Media.

Govett, M., Bah, B., Bauer, P., et al. (2024) Exascale Computing and Data Handling: Challenges and Opportunities for Weather and Climate Prediction. *Bulletin of the American Meteorological Society.*

Graham, S. S., Harrison, K. R., Edward, et al. (2024). Beyond bias: Aggregate approaches to conflicts of interest research and policy in biomedical research. *World Medical & Health Policy.*

Granata, G., Astorri, R., Broens, E. M., et al. (2024) The World Health Organization Pandemic Agreement draft: Considerations by the ESCMID Emerging Infections Task Force (EITaF). *Clinical Microbiology and Infection.*

Greer, S. L., Fonseca, E. M., Raj, M., et al. (2024) Institutions and the politics of agency in COVID-19 response: Federalism, executive power, and public health policy in Brazil, India, and the US. *Journal of Social Policy* 53(3): 792-810.

Grinin, L., Korotayev, A. (2024) Is the Fifth Generation of Revolution Studies Still Coming?

Gross, F. (2019) Occam's razor in molecular and systems biology. *Philosophy of Science* 86(5): 1134-1145.

Gundersen, T. (2024) Trustworthy science advice: The case of policy recommendations. *Res Publica* 30(1): 125-143.

Gustian, D., Marzuki, M., Nukman, N., et al. (2024) Synergies in Education: Integrating Character, Literacy, And Technology for Enhanced Outcomes: Current Perspectives from Global Education Experts. *International Journal of Teaching and Learning* 2(2): 498-512.

Hamid, A. R. A. H. (2020). Social responsibility of medical journal: a concern for COVID-19 pandemic. *Medical Journal of Indonesia* 29(1): 1-3.

Hanisch, S., Eirdosh, D. (2023) Teaching for the interdisciplinary understanding of evolutionary concepts. *In* Evolutionary Thinking Across Disciplines: Problems and Perspectives in Generalized Darwinism. Cham: Springer International Publishing.

Hanna, R. (2024) *Science for Humans: Mind, Life, the Formal-& Natural Sciences, and a New Concept of Nature.* Springer Nature.

Hansson, S. O. (2019) Defining pseudoscience and science. *Philosophy of Pseudoscience: Reconsidering the demarcation problem.* University of Chicago Press: Chicago, IL.

Hansson, S. O. (2024) Scientific Expertise is Needed to Identify Pseudoscience. *Acta Baltica Historiae et Philosophiae Scientiarum* 12(1): 117-127.

Hardwicke, T. E., Wagenmakers, E. J. (2023) Reducing bias, increasing transparency and calibrating confidence with preregistration. *Nature Human Behaviour* 7(1): 15-26.

Hartle, J. B. (2021) *Gravity: an introduction to Einstein's general relativity.* Cambridge University Press.

Haselton, M. G., Nettle, D., Andrews, P. W. (2015) The evolution of cognitive bias. *The Handbook of Evolutionary Psychology.*

Haslam, S. A., McGarty, C., Cruwys, T., et al. (2024) *Research methods and statistics in psychology.* SAGE Publications Limited.

Hicks, D. J. (2023) Open science, the replication crisis, and environmental public health. *Accountability in Research* 30(1): 34-62.

Higgins, K. M., Levin, G., Busch, R. (2024) Considerations for open-label randomized clinical trials: Design, conduct, and analysis. *Clinical Trials.*

Hillersdal, L., Jespersen, A. P., Oxlund, B., et al. (2020). Affect and effect in interdisciplinary research collaboration. *Science & Technology Studies* 33(2): 66-82.

Hinterleitner, M., Knill, C., Steinebach, Y. (2024) The growth of policies, rules, and regulations: A review of the literature and research agenda. *Regulation & Governance* 18(2):637-654.

Hirose, M., Creswell, J. W. (2023) Applying core quality criteria of mixed methods research to an empirical study. *Journal of Mixed Methods Research* 17(1): 12-28.

Holm, P., Goodsite, M. E., Cloetingh, et al. (2013). Collaboration between the natural, social and human sciences in global change research. *Environmental Science & Policy* 28: 25-35.

Hooker, C. (2022) Understanding HPS paradigms through Galison's problems. *Axiomathes* 32(6): 931-956.

Hourdequin, M. (2024) *Environmental ethics: From theory to practice.* Bloomsbury Publishing.

Huang, J. (2024) Diagnosing the declining industry sponsorship in clinical research. *Scientometrics* 129(1): 663-679.

Huang, L. K., Chao, S. P., Hu, C. J. (2020) Clinical trials of new drugs for Alzheimer disease. *Journal of Biomedical Science* 27: 1-13.

Huntington-Klein, N. (2021) *The effect: An introduction to research design and causality.* Chapman and Hall/CRC.

Institute of Medicine. 1992. Responsible Science: Ensuring the Integrity of the Research Process: Volume I. Washington, DC: The National Academies Press.

Institute of Medicine. 1993. Responsible Science: Ensuring the Integrity of the Research Process: Volume II. Washington, DC: The National Academies Press.

Intemann, K. (2023) Science communication and public trust in science. *Interdisciplinary Science Reviews* 48(2): 350-365.

InterAcademy Partnership. (2016). *Doing global science: A guide to responsible conduct in the global research enterprise.* Princeton University Press.

Iseman, T., Miralles-Wilhelm, F. (2021) *Nature-based solutions in agriculture: The case and pathway for adoption.* Food & Agriculture Org.

Jaeger, J., Masselot, C., Greshake Tzovaras, B., et al. (2023) An epistemology for democratic citizen science. *Royal Society Open Science* 10(11): 231100.

Jafari, M. (2024) Ethics and Pseudoscience in Our Daily Lives: The Status of the Science Behind Dietary Supplements. *How Science Engages with Ethics and Why It Should: An Interdisciplinary Approach.*

Jain, A. S., Shelke, S. S., Jadhav, V., et al. V. (2024) An Overview of Data Integrity in Pharmaceutical Industry. *African J Biological Sciences* 6(12).

Jalal, A. (2024) The Power of Multidisciplinary Collaboration in Scientific Research. *Kashf Journal of Multidisciplinary Research* 1(04): 176-185.

Jana, S. (2019) A history and development of peer-review process. *Annals of Library and Information Studies* 66(4), 152-162.

Jansson, L. (2021) The Explanatory Value of Selecting the Appropriate Scale (S). *In* The Routledge Companion to Philosophy of Physics. Routledge.

Jarvis, M. F. (2024). Decatastrophizing research irreproducibility. *Biochemical Pharmacology.*

Jerome, L. W., Paterson, S. K., von Stamm, B., et al. (2024) Making Transdisciplinarity Work for Complex Systems: A Dynamic Model for Blending Diverse Knowledges. *Futures* 103415.

Jonassen, D. H., Hung, W. (2015). All problems are not equal: Implications for problem-based learning. *In* Essential readings in problem-based learning: Exploring and extending the legacy of Howard S. Barrows.

Kaiho, K. (2023). An animal crisis caused by pollution, deforestation, and warming in the late 21st century and exacerbation by nuclear war. *Heliyon* 9(4).

Kapon, S., Schvartzer, M. (2024). Guided inquiry into a physics equation. *Cognition and Instruction* 42(1): 159-206.

Kavanagh, C., Agan, L., Sneider, C. (2005) Learning about phases of the moon and eclipses: A guide for teachers and curriculum developers. *Astronomy Education Review* 4(1): 19-52.

Kennedy Jr, R. F. (2023) *Limited boxed set: The real Anthony Fauci: Bill Gates, big pharma, and the global war on democracy and public health*. Simon and Schuster.

Kepes, S., Wang, W., Cortina, J. M. (2023) Assessing publication bias: A 7-step user's guide with best-practice recommendations. *Journal of Business and Psychology* 38(5): 957-982.

Khan, R., Usman, M., Moinuddin, M. (2024) The big data revolution: Leveraging vast information for competitive advantage. *Revista Espanola de Documentacion Científica* 18(02): 65-94.

Khang, A., Hajimahmud, A. V., Triwiyanto, T., et al. (2024). Cloud Platform and Data Storage Systems in the Healthcare Ecosystem. *In Medical Robotics and AI-Assisted Diagnostics for a High-Tech Healthcare Industry*. IGI Global.

Kirillova, E. (2020, April) The role of scientific and industrial cooperation in assessing the innovative potential of an industrial enterprise and the approach to evaluation through joint patent and licensing activities. *In* Proceeding of the International Science and Technology Conference "FarEastCon 201" October 2019, Vladivostok, Russian Federation, Far Eastern Federal University. Singapore: Springer Singapore.

Kirschbaum, K. (2024) Climate Change Education for Student Agency. *School of Education and Leadership Student Capstone Projects* 1056.

Kishan, S., Gupta, U. Research Publication Ethics: Ensuring Integrity, Transparency, and Responsible Scholarship. *J Tech Education*.

Kitmacher, G. H., Gerstenmaier, W. H., Bartoe, et al. (2005) The international space station: A pathway to the future. *Acta Astronautica* 57(2-8): 594-603.

Klonsky, E. D. (2024) Campbell's Law Explains the Replication Crisis: Pre-Registration Badges Are History Repeating. *Assessment*.

Kneer, M., Skoczeń, I. (2023) Outcome effects, moral luck and the hindsight bias. *Cognition* 232: 105258.

Komalasari, R., Mustafa, C. (2024) Navigating governance for equitable health systems: Ethical dimensions, ideological impacts, and evidence-based improvements. *Governance* 12(1): 49-62.

Koonin, S. E. (2024) *Unsettled (Updated and Expanded Edition): What Climate Science Tells Us, What It Doesn't, and Why It Matters.* BenBella Books.

Krakowski, A., Greenwald, E., Roman, N., et al. (2024). Computational Thinking for Science: Positioning coding as a tool for doing science. *Journal of Research in Science Teaching* 61(7): 1574-1608.

Kraus, S., Jones, P., Kailer, N., et al. (2021) Digital transformation: An overview of the current state of the art of research. *Sage Open* 11(3): 21582440211047576.

Kretser, A., Murphy, D., Bertuzzi, S., et al. (2019) Scientific integrity principles and best practices: recommendations from a scientific integrity consortium. *Science and Engineering Ethics* 25: 327-355.

Krishna, V. V. (2024) *The Indian Science Community: Historical and Sociological Studies.* Taylor & Francis.

Krychtiuk, K. A., Andersson, T. L., Bodesheim, U., et al. (2024) Drug development for major chronic health conditions—aligning with growing public health needs: Proceedings from a multistakeholder think tank. *American Heart Journal* 270: 23-43.

Kuhn, H., Waldeck, D. H., Försterling, H. D. (2024) *Principles of physical chemistry.* John Wiley & Sons.

Kuhn, TS. (1962) *The Structure of Scientific Revolutions.* University of Chicago Press: Chicago, IL.

Kumar, S. (2024) Environmental sustainability. *In Green Transition Impacts on the Economy, Society, and Environment.*

Kumar, S., Verma, A. K., Mirza, A. (2024). Artificial Intelligence and Climate Change Mitigation. *In* Digital Transformation, Artificial Intelligence and Society: Opportunities and Challenges. Singapore: Springer Nature Singapore.

Ladner, D. P., Goldstein, A. M., Billiar, et al. (2024). Transforming the Future of Surgeon-Scientists. *Annals of Surgery* 279(2): 231-239.

Lambright, W. H. (2010) Exploring space: NASA at 50 and beyond. *Public Administration Review* 70(1): 151-157.

Langhammer, P. F., Bull, J. W., Bicknell, J. E., et al. (2024). The positive impact of conservation action. *Science* 384(6694): 453-458.

Larson, E. J. (2006) *Evolution: The remarkable history of a scientific theory* (Vol. 17). Modern Library.

Latour, B, Woolgar, S. (1979) *Laboratory Life: The Construction of Scientific Facts.* Princeton University Press: Princeton, NJ.

Laudan, L. (2020) The history of science and the philosophy of science. *In* Companion to the history of modern science. Routledge.

Lee, M. (2024) Role of Bioinformatics in Precision Oncology: Analyzing Big Data for Personalized Treatment. *Revista de Inteligencia Artificial en Medicina* 15(1): 221-231.

Leichenko, R., O'Brien, K. (2024) *Climate and society: Transforming the future.* John Wiley & Sons.

Leiss, W. (2023) *The domination of nature: New edition.* McGill-Queen's Press-MQUP.

Leś, T. (2024) Education, ideology, and critical thinking. *In* Defending the Value of Education as a Public Good. Routledge.

Lipinski, T. A., Lee, J. (2024) Collection Development and Maintenance of Accurate Grey Literature on Climate Change: A Case Study of the Law and Policy in the United States. *The Grey Journal* 20(2).

Li, T., Yue, X. G., Qin, M., et al. (2024) Towards Paris Climate Agreement goals: The essential role of green finance and green technology. *Energy Economics* 129: 107273.

Li, X., Cai, F., Bao, J., et al. (2024) Navigating interdisciplinary research: Historical progression and contemporary challenges. *Journal of Data and Information Science* 9(3): 14-28.

Lin, Y. K., Maruping, L. M. (2022). Open-source collaboration in digital entrepreneurship. *Organization Science* 33(1): 212-230.

Lin, Y., Yang, W., Zhang, H., et al. (2024) Return to the Moon: New Perspectives on Lunar Exploration. *Science Bulletin.*

Little, J. C., Kaaronen, R. O., Hukkinen, J. I., et al. (2023). Earth Systems to Anthropocene Systems: An Evolutionary, System-of-Systems, Convergence Paradigm for Interdependent Societal Challenges. *Environmental Science & Technology* 57(14): 5504-5520.

Liu, G., Lin, Q., Jin, S., et al. (2022) The CRISPR-Cas toolbox and gene editing technologies. *Molecular Cell* 82(2): 333-347.

Liu, L., Jones, B. F., Uzzi, B., et al. (2023) Data, measurement and empirical methods in the science of science. *Nature Human Behaviour* 7(7): 1046-1058.

Liu, Y. (2024). A time-series analysis of the evolution of scientific and technological concepts in the change of social thinking in modern China. *Applied Mathematics and Nonlinear Sciences* 9(1).

Longdon, J., Gabrys, J., Blackwell, A. F. (2024) Taking data science into the forest. *Interdisciplinary Science Reviews* 49(1): 82-103.

Lövestam, G., Bremer-Hoffmann, S., Jonkers, K., et al (2024). Fostering scientific integrity and research ethics in a science-for-policy research organisation. *Research Ethics*.

Lu, D. (2024). *Regional development and its spatial structure*. Springer.

Lupia, A., Allison, D. B., Jamieson, K. H., et al. (2024) Trends in US public confidence in science and opportunities for progress. *Proceedings of the National Academy of Sciences* 121(11): e2319488121.

Lynn, S. J., Aksen, D., Sleight, F., et al. (2023). Combating pseudoscience in clinical psychology: From the scientific mindset, to busting myths, to prescriptive remedies. *In* Toward a science of clinical psychology: A tribute to the life and works of Scott O. Lilienfeld. Cham: Springer International Publishing.

Macrina, F. L. (2014) *Scientific integrity: Text and cases in responsible conduct of research*. John Wiley & Sons.

Manoj, K. M., Jaeken, L. (2023) Synthesis of theories on cellular powering, coherence, homeostasis and electro-mechanics: Murburn concept and evolutionary perspectives. *Journal of Cellular Physiology* 238(5): 931-953.

Masters, L. (2024) A Diplomatic Conduit: The Role of Science Diplomacy in Africa. *In* Key Issues in African Diplomacy: Developments and Achievements.

Matthews, M. R. (2009). Science, worldviews and education: An introduction. *Science & Education* 18: 641-666.

Matthews, M. R. (2024) Thomas Kuhn and science education: Learning from the past and the importance of history and philosophy of science. *Science & Education* 33(3): 609-678.

Mayr, E. (1997) *This Is Biology: The Science of the Living World.* Harvard University Press: Cambridge, MA.

McElwain, M. W., Feinberg, L. D., Perrin, M. D., et al. (2023). The James Webb Space Telescope mission: optical telescope element design, development, and performance. *Publications of the Astronomical Society of the Pacific* 135(1047): 058001.

Mercan, B. (2024) *Preservice science teachers' system thinking skills and conceptual understanding for water cycle.* [M.S. - Master of Science]. Middle East Technical University.

Mietchen, D., Jeschke, J., Heger, T. (2024) Introducing Hypothesis Descriptions. *Research Ideas and Outcomes* 10: e119805.

Miralles-Wilhelm, F. (2021) *Nature-based solutions in agriculture: Sustainable management and conservation of land, water and biodiversity.* Food & Agriculture Org.

Mocanu, M., Rusu, V. D., Bibiri, A. D. (2024) Competing for Research Funding: Key Elements Impacting the Evaluation of Grant Proposal. *Heliyon.*

Molthan-Hill, P., Blaj-Ward, L., Mbah, M. F., et al. (2022) Climate change education at universities: Relevance and Strategies for Every Discipline. *In* Handbook of Climate Change Mitigation and Adaptation. Cham: Springer International Publishing.

Moneshwaran, S., Macrin, D., Kanagathara, N. (2024) An unprecedented global challenge, emerging trends and innovations in the fight against COVID-19: A comprehensive review. *International Journal of Biological Macromolecules.*

Moomaw, S. (2024) *Teaching STEM in the early years: Activities for integrating science, technology, engineering, and mathematics.* Redleaf Press.

Moore, John A. (1993) *Science as a way of knowing. The foundations of modern biology.* Harvard University Press: Cambridge, MA.

Moser, K. (2024) Climate Change Denial: An Ecocidal, Parallel Universe of Simulation. *In* Fake News in Contemporary Science and Politics: A Requiem for the Real? Cham: Springer Nature Switzerland.

Moşteanu, N. R. (2024) Adapting to the Unpredictable: Building Resilience for Business Continuity in an Ever-Changing Landscape. *European Journal of Theoretical and Applied Sciences* 2(1): 444-457.

Munari, F., Leonardelli, E., Menini, S., et al. (2024) Public research funding and science-based innovation: An analysis of ERC research grants, publications and patents. *Research Evaluation*.

Nałęcz-Charkiewicz, K., Charkiewicz, K., Nowak, R. M. (2024) Quantum computing in bioinformatics: a systematic review mapping. *Briefings in Bioinformatics* 25(5): bbae391.

Namdarian, L., Khedmatgozar, H. R. (2024) How does the evidence-based approach contribute to ethical policy making in science and technology? *Iranian Journal of Information Processing and Management* 39(Special Issue 2): 1-26.

Nanglu, K., de Carle, D., Cullen, T. M., Anderson, et al. (2023) The nature of science: The fundamental role of natural history in ecology, evolution, conservation, and education. *Ecology and Evolution* 13(10): e10621.

National Research Council, Division on Earth, Life Studies, Board on Life Sciences, & Committee on A Framework for Developing a New Taxonomy of Disease. (2011) Toward precision medicine: building a knowledge network for biomedical research and a new taxonomy of disease.

Nature of Science-Position Paper. (2022) National Science Teachers Association.

Nayarisseri, A., Khandelwal, R., Tanwar, P., et al. (2021) Artificial intelligence, big data and machine learning approaches in precision medicine & drug discovery. *Current Drug Targets* 22(6): 631-655.

Newman, J. (2024) Promoting interdisciplinary research collaboration: A Systematic Review, a critical literature Review, and a pathway forward. *Social Epistemology* 38(2): 135-151.

Ng, Y. H. (2023) Questioning authority through a scientific inquiry on heliocentrism. *Physics Education* 58(3): 035009.

NGSS Lead States. (2013) Next Generation Science Standards. National Academies Press.

Nicholl, D. S. (2023) *An introduction to genetic engineering.* Cambridge University Press.

Nielsen, M. (2020) *Reinventing discovery: the new era of networked science.* Princeton University Press: Princeton, NJ.

Nightingale, D. J., Rhodes, D. H. (2024). *Architecting the future enterprise.* MIT Press.

Nosek, B. A., Errington, T. M. (2020) What is replication? *PLoS Biology* 18(3): e3000691.

Novelli, E., Spina, C. (2024) How do entrepreneurs benefit from acting like scientists in business model development? Strategic commitments, uncertainty and economic performance. *Strategic Management Journal.*

NSTA. (2021) National Science Teaching Association Position Statement: Nature of Science.

Nunes, C., McKee, M., Howard, N. (2024) The role of global health partnerships in vaccine equity: A scoping review. *PLOS Global Public Health* 4(2): e0002834.

Nunn, R., Brandt, C., Deveci, T. (2018) Transparency, subjectivity and objectivity in academic texts. *ESBB English Scholarship Beyond Borders* 4(1): 71-102.

Nuzzo, R. (2014) Scientific method: statistical errors. *Nature* 506 (7487).

Olaniyi, O. O., Ugonnia, J. C., Olaniyi, F. G., et al. (2024). Digital collaborative tools, strategic communication, and social capital: Unveiling the impact of digital transformation on organizational dynamics. *Asian Journal of Research in Computer Science* 17(5):140-156.

Oliveira, A., Feyzi Behnagh, R., Ni, L., et al. (2019) Emerging technologies as pedagogical tools for teaching and learning science: A literature review. *Human Behavior and Emerging Technologies* 1(2): 149-160.

Omilani, N. A. (2024) Integrating indigenous knowledge into chemistry and science education. *International Journal of studies in Psychology* 4(1): 55-58.

Oruh, S., Agustang, A. (2024) The Contribution of the Philosophy of Science to Scientific Research and Social Life. *International Journal of Health, Economics, and Social Sciences (IJHESS)* 6(1): 220-228.

Osasona, F., Amoo, O. O., Atadoga, A., et al. (2024) Reviewing the ethical implications of AI in decision making processes. *International Journal of Management & Entrepreneurship Research* 6(2): 322-335.

Osborne, J. (2023) Science, scientific literacy, and science education. *In* Handbook of research on science education. Routledge.

Oschman, J. L. (2015) *Energy medicine: The scientific basis.* Elsevier Health Sciences.

Oswald, T. K., Rumbold, A. R., Kedzior, et al. (2020). Psychological impacts of "screen time" and "green time" for children and adolescents: A systematic scoping review. *PloS One* 15(9): e0237725.

Our Definition of Science. (2024) The Science Council.

Pajo, B. (2022). *Introduction to research methods: A hands-on approach.* Sage publications.

Pallett, H. (2020) The new evidence-based policy: public participation between 'hard evidence' and democracy in practice. *Evidence & Policy* 16(2): 209-227.

Pandey, S., Patel, A. (2024) Leveraging ChatGPT in Law Enforcement. *Recent Advances in Computer Science and Communications* 17(2): 40-48.

Papale, F., Doolittle, W. F. (2024) Towards a More General Theory of Evolution by Natural Selection: A Manifesto. *In* Philosophy, Theory, and Practice in Biology.

Pardeshi, S., Dhodapkar, R., Namdeo, S. K., et al. (2024). *Science Diplomacy.*

Park, A., Maine, E., Fini, R., et al. (2024) Science-based innovation via university spin-offs: the influence of intangible assets. *R&D Management* 54(1): 178-198.

Park, S. J., Kim, Y. Y., Han, J. Y., et al. (2024) Advancements in Human Embryonic Stem Cell Research: Clinical Applications and Ethical Issues. *Tissue Engineering and Regenerative Medicine* 21(3): 379-394.

Parker, O. N., Short, C. E., Titus, et al. (2024) Accentuate the positive? Strategic negativity amid the hazard of high expectations. *Strategic Management Journal.*

Parolin, G., Mcaloone, T. C., Pigosso, D. C. (2024) How can technology assessment tools support sustainable innovation? A systematic literature review and synthesis. *Technovation* 129: 102881.

Pasternak-Taschner, N., Almeida, P. (2024) Teaching scientific evidence and critical thinking for policy making. *Biology Methods and Protocols.*

Patel, T. A., Jain, B., Raifman, J. (2024) Revamping public health systems: Lessons learned from the tripledemic. *American Journal of Preventive Medicine* 66(1): 185-188.

Pattison, A. J., Pedroso, C., Cohen, B. E., et al. (2024). Automated High-Resolution Phase-Contrast Scanning Transmission Electron Microscopy. *Microscopy and Microanalysis* 30(Supplement_1).

Patton, MQ. (2018) Evaluation Science. *American Journal of Evaluation* 39(2): 183-200.

Pearl, J. (2020) The art and science of cause and effect. *In* Shaping Entrepreneurship Research. Routledge.

Peeples, J., Murphy, M. (2022) Discourse and rhetorical analysis approaches to environment, media, and communication. *In* The Routledge handbook of environment and communication. Routledge.

Peng, R. D., Hicks, S. C. (2021) Reproducible research: a retrospective. *Annual Review of Public Health* 42(1): 79-93.

Pennisi, F., Genovese, C., Gianfredi, V. (2024). Lessons from the COVID-19 Pandemic: Promoting Vaccination and Public Health Resilience, a Narrative Review. *Vaccines* 12(8): 891.

Pérez-Bentancur, V., Tiscornia, L. (2024) Iteration in mixed-methods research designs combining experiments and fieldwork. *Sociological Methods & Research* 53(2): 729-759.

Perone, G. (2020) An ARIMA Model to Forecast the Spread and the Final Size of COVID-2019 Epidemic in Italy. Health Econometrics and Data Group Working Paper Series, University of York.

Periyasamy, G., Gupta, H., Chatterjee, S., et al. (2024) Expounding Conflicts of Interest. *In* Scientific Publishing Ecosystem: An Author-Editor-Reviewer Axis. Singapore: Springer Nature Singapore.

Petersen, I. T., Apfelbaum, K. S., McMurray, B. (2024) Adapting open science and pre-registration to longitudinal research. *Infant and Child Development* 33(1): e2315.

Peterson, D., Panofsky, A. (2021) Self-correction in science: The diagnostic and integrative motives for replication. *Social Studies of Science* 51(4): 583-605.

Pezzullo, A. M., Ioannidis, J. P., Boccia, S. (2023) Quality, integrity and utility of COVID-19 science: opportunities for public health researchers. *European Journal of Public Health* 33(2): 157-158.

Plebanski, J., Krasinski, A. (2024) *An introduction to general relativity and cosmology*. Cambridge University Press.

Poincaré, H. (2022) *The foundations of science: Science and hypothesis, the value of science, science and method*. DigiCat.

Pols, J., M'charek, A., Jerak-Zuiderent, S., et al. (2024) Achieving good science: The integrity of scientific institutions. *Learning and Teaching* 17(1): 24-53.

Popper, K. (1959) *The Logic of Scientific Discovery*. Routeledge Classics: New York, NY.

Porsdam, H., Porsdam Mann, S. (2024) Anticipation and diplomacy (with) in science: activating the right to science for science diplomacy. *The International Journal of Human Rights* 28(3): 480-496.

Post, S., Bienzeisler, N. (2024) The Honest Broker versus the Epistocrat: Attenuating Distrust in Science by Disentangling Science from Politics. *Political Communication*.

Pradeu, T., Lemoine, M., Khelfaoui, M., et al. (2024) Philosophy in Science: Can philosophers of science permeate through science and produce scientific knowledge? *The British Journal for the Philosophy of Science* 75(2).

Prasad, V., Haslam, A. (2024) COVID-19 vaccines: history of the pandemic's great scientific success and flawed policy implementation. *Monash Bioethics Review*.

Psillos, S. (2008) Philosophy of science. *In* The Routledge Companion to Twentieth Century Philosophy. Routledge.

Puaschunder, J. M. (2024) A science diplomacy agenda for sustainable development research. *In* An Agenda for Sustainable Development Research. Cham: Springer Nature Switzerland.

Pugh, J., Savulescu, J., Brown, R. C., et al. (2024) The ethics of natural immunity exemptions to vaccine mandates: the Supreme Court petition. *Journal of Medical Ethics*.

Quinn, A. (2021) Transparency and secrecy in citizen science: Lessons from herping. *Studies in History and Philosophy of Science Part A* 85: 208-217.

Quoc, N. A., Linh, L. N. (2022) The Scientific Essence. *International Journal of Social Science and Human Research* 5(10): 2644-0679.

Rådberg, K. K., Löfsten, H. (2024) The entrepreneurial university and development of large-scale research infrastructure: Exploring the emerging university function of collaboration and leadership. *The Journal of Technology Transfer* 49(1): 334-366.

Rader, J. A., Hedrick, T. L. (2023) Morphological evolution of bird wings follows a mechanical sensitivity gradient determined by the aerodynamics of flapping flight. *Nature Communications* 14(1): 7494.

Rafiee Rad, S., Braun, S. T., Roy, O. (2024) Anchoring as a Structural Bias of Deliberation.

Rahman, A., Shah, M., Shord, S. S. (2024) Dosage Optimization: A Regulatory Perspective for Developing Oncology Drugs. *Clinical Pharmacology & Therapeutics*.

Ramos-Vielba, I., Bloch, C., Thomas, D. A., et al. (2024) How can societally-targeted research funding shape researcher networks and practices? *Research Evaluation*.

Ramstead, M. J., Sakthivadivel, D. A., Heins, et al. (2023). On Bayesian mechanics: a physics of and by beliefs. *Interface Focus* 13(3): 20220029.

Razavi, B. (2021) *Fundamentals of microelectronics.* John Wiley & Sons.

Reichle, D. E. (2023) *The global carbon cycle and climate change: Scaling ecological energetics from organism to the biosphere.* Elsevier.

Reneau, A., & Reneau, A. (2021) Moon and Mars: A Comparison of the Two. *In* Moon First and Mars Second: A Practical Approach to Human Space Exploration.

Resnik, D. B., Elliott, K. C. (2023) Science, values, and the new demarcation problem. *Journal for General Philosophy of Science* 54(2): 259-286.

Restivo, S., Loughlin, J. (2000) The invention of science. *Cultural Dynamics* 12(2): 135-149.

Ripple, K. J., Hudson, C., Knight, E., et al. (2024) Enabling usable science takes a community: Using our roles as funders to catalyze change. *PLoS Biology* 22(6): e3002675.

Ristori, M. V., Guarrasi, V., Soda, P., et al. (2024) Emerging Microorganisms and Infectious Diseases: One Health Approach for Health Shared Vision. *Genes* 15(7).

Ritoré, Á., Jiménez, C. M., González, J. L., et al. (2024) The role of Open Access Data in democratizing healthcare AI: A pathway to research enhancement, patient well-being and treatment equity in Andalusia, Spain. *PLOS Digital Health* 3(9): e0000599.

Rockström, J., Kotzé, L., Milutinović, S., et al. (2024). The planetary commons: A new paradigm for safeguarding Earth-regulating systems in the Anthropocene. *Proceedings of the National Academy of Sciences* 121(5): e2301531121.

Rodney, S. (2024) Historical Methods. *In* An Introductory Guide to Qualitative Research in Art Museums. Routledge.

Rogers, B. A., Chicas, H., Kelly, J. M., et al. (2023). Seeing your life story as a Hero's Journey increases meaning in life. *Journal of Personality and Social Psychology.*

Roje, R., Reyes Elizondo, A., Kaltenbrunner, W., et al. (2023) Factors influencing the promotion and implementation of research integrity in research performing and research funding organizations: A scoping review. *Accountability in Research* 30(8): 633-671.

Romanova, A., Rubinelli, S., Diviani, N. (2024) Improving health and scientific literacy in disadvantaged groups: A scoping review of interventions. *Patient Education and Counseling*.

Rosenberg, A., McIntyre, L. (2019) *Philosophy of science: A contemporary introduction*. Routledge.

Rosman, T., Grösser, S. (2024) Belief updating when confronted with scientific evidence: Examining the role of trust in science. *Public Understanding of Science* 33(3): 308-324.

Rossoni, A. L., de Vasconcellos, E. P. G., de Castilho Rossoni, R. L. (2024) Barriers and facilitators of university-industry collaboration for research, development and innovation: a systematic review. *Management Review Quarterly* 74(3): 1841-1877.

Rouse, J. (2018) *Engaging science: How to understand its practices philosophically*. Cornell University Press.

Roy, A., Newman, A., Round, H., et al. (2024) Ethical culture in organizations: A review and agenda for future research. *Business Ethics Quarterly* 34(1): 97-138.

Rudolph, J. L. (2024) Scientific literacy: Its real origin story and functional role in American education. *Journal of Research in Science Teaching* 61(3): 519-532.

Rueda, J. (2024) Genetic enhancement, human extinction, and the best interests of posthumanity. *Bioethics* 38(6): 529-538.

Rutjens, B. T., Van der Linden, S., Van der Lee, R. (2021) Science skepticism in times of COVID-19. *Group Processes & Intergroup Relations* 24(2): 276-283.

Ruxton, C. (2022) Interpretation of observational studies: the good, the bad and the sensational. *Proceedings of the Nutrition Society* 81(4): 279-287.

Ryff, C. D. (2018) Well-being with soul: Science in pursuit of human potential. *Perspectives on Psychological Science* 13(2): 242-248.

Sagan, C. (1995) *The Demon-Haunted World: Science as a Candle in the Dark.* Random House Publishing: New York.

Sakurai, J. J., Napolitano, J. (2020) *Modern quantum mechanics.* Cambridge University Press.

Saldana, P. M. (2024) *The Influence of Pandemic Financial Relief on Organizational Development and Business Continuity* (Doctoral dissertation, California Baptist University).

Samuel-Okon, A. D., Abejide, O. O. (2024) Bridging the digital divide: Exploring the role of artificial intelligence and automation in enhancing connectivity in developing nations. *Journal of Engineering Research and Reports* 26(6): 165-177.

Sandberg, J., Alvesson, M. (2021). Meanings of theory: Clarifying theory through typification. *Journal of Management Studies* 58(2): 487-516.

Santos, L. F. (2017) The role of critical thinking in science education. *Online Submission* 8(20): 160-173.

Saranya, S., Thamanna, L., Chellapandi, P. (2024) Unveiling the potential of systems biology in biotechnology and biomedical research. *Systems Microbiology and Biomanufacturing.*

Savaresi, A. (2016). The Paris Agreement: a new beginning? *Journal of Energy & Natural Resources Law* 34(1): 16-26.

Scherrer, R. J. (2024) *Quantum mechanics: an accessible introduction.* World Scientific.

Schickore, J. (2011) More thoughts on HPS: Another 20 years later. *Perspectives on Science* 19(4): 453-481.

Schmidt, M. H., Dluzen, D. F. (2024) Rigor and reproducibility in genetic research and the effects on scientific reporting and public discourse. *In* Rigor and Reproducibility in Genetics and Genomics. Academic Press.

Schubert, P. (2024) IRECS Framework: Identification of Requirements for Enterprise Collaboration Systems. *Procedia Computer Science* 239: 1467-1475.

Schulz, P. J., Nakamoto, K. (2024) Worse Than Ignorance: The Challenge of Health Misinformation. *Elements in Health Communication.*

Science. (2024) Merriam-Webster.com Dictionary. @ https://www.merriam-webster.com/dictionary/science.

Science. (2024) Oxford Reference. Oxford, England: Oxford University Press.

Scotese, C. R., Vérard, C., Burgener, L., et al. (2025) The Cretaceous World: Plate Tectonics, Paleogeography, and Paleoclimate. *Geological Society, London, Special Publications.* 544(1): SP544-2024.

Szczuka, J. M., Meinert, J., Krämer, N. C. (2024) Listen to the Scientists: Effects of Exposure to Scientists and General Media Consumption on Cognitive, Affective, and Behavioral Mechanisms During the COVID-19 Pandemic. *Human Behavior and Emerging Technologies* 2024(1): 8826396.

Sears, N. A. (2020) Existential security: Towards a security framework for the survival of humanity. *Global Policy* 11(2): 255-266.

Seethaler, S. L. (2024) *Beyond the Sage on the Stage: Communicating Science and Contemporary Issues Effectively.* University of Toronto Press.

Serbe-Kamp, É., Bemme, J., Pollak, D., et al. (2023) Open Citizen Science: fostering open knowledge with participation. *Research Ideas and Outcomes* 9: e96476.

Sessler, D. I., Alman, B., Treggiari, M. M., et al. (2023) Pro-con debate: interdisciplinary perspectives on industry-sponsored research. *Anesthesia & Analgesia* 136(6): 1055-1063.

Sherani, A. M. K., Khan, M., Qayyum, M. U., et al. (2024) Synergizing AI and Healthcare: Pioneering Advances in Cancer Medicine for Personalized Treatment. *International Journal of Multidisciplinary Sciences and Arts* 3(1): 270-277.

Shelton, R. C., Brownson, R. C. (2024). Enhancing impact: a call to action for equitable implementation science. *Prevention Science* 25(Suppl 1): 174-189.

Shen, C. H. (2023) *Diagnostic Molecular Biology.* Elsevier.

Shorina, T., Abysova, M., Poda, T., et al. (2024). A holistic approach to addressing the global environmental challenge: the scientific-philosophical methodology. *Amazonia Investigation* 13(75): 45-55.

Sikorski, M. (2024) Values, bias and replicability. *Synthese* 203(5): 1-25.

Symons, J., Dixon, T. A., Dalziell, J., et al. (2024) Engineering biology and climate change mitigation: Policy considerations. *Nature Communications* 15(1): 2669.

Slater, W. H. (2023) Rhetoric and the Stases: A Universal Critical Thinking Problem-Solving Framework for the Sciences and Arts. *In* Brain, Decision Making and Mental Health. Cham: Springer International Publishing.

Smedberg, A. (2021) Methods matter: Anecdotalisation as knowledge co-creation. *Urban Matters* 1(1).

Smillie, L., Scharfbillig, M. (2024) Trustworthy Public Communications, Publications Office of the European Union, Luxembourg; accessed @https://data.europa.eu/doi/10.2760/695605.

So, A. (2024) The implications of ethical perspectives in AI and autonomous systems. *In* Ethics in Online AI-based Systems. Academic Press.

Sober, E. (2024) *The Philosophy of Evolutionary Theory: Concepts, Inferences, and Probabilities.* Cambridge University Press.

Soloman, S. (2024) The Truth of the Origin of the Universe. *In* The Truth of the Origin of the Universe: The Universe, The Origin of Life in the Universe: The Black Hole and God. Khanna Publishing House.

Souchet, B., Michaïl, A., Billoir, B., et al. (2023) Biological Diagnosis of Alzheimer's Disease Based on Amyloid Status: An Illustration of Confirmation Bias in Medical Research? *International Journal of Molecular Sciences* 24(24): 17544.

Souza, D. V. L. D., Oliveira, I. M. D. (2024) Pseudosciences and the Current Challenges Imposed on Science Teaching. *Educação & Realidade* 49: e121157.

Stewart Jr, C. N. (2023) *Research ethics for scientists: A companion for students.* John Wiley & Sons.

Stoner, J. L., Felix, R., Stadler Blank, A. (2023) Best practices for implementing experimental research methods. *International Journal of Consumer Studies* 47(4): 1579-1595.

Storm, H., Baylis, K., Heckelei, T. (2020) Machine learning in agricultural and applied economics. *European Review of Agricultural Economics* 47(3): 849-892.

Stoudt, S., Jernite, Y., Marshall, B., et al. (2024). Ten simple rules for building and maintaining a responsible data science workflow. *PLOS Computational Biology* 20(7): e1012232.

Straf, M. L., Schwandt, T. A., Prewitt, K. (Eds.) (2012) *Using science as evidence in public policy*. National Academies Press.

Subbiah, V. (2023) The next generation of evidence-based medicine. *Nature Medicine* 29(1): 49-58.

Suiter, S., Byars-Winston, A., Sancheznieto, F., et al. (2024). Utilizing mentorship education to promote a culturally responsive research training environment in the biomedical sciences. *Plos One* 19(8): e0291221.

Sultan, F. (2024) The Future of Multidisciplinary Research: Trends and Opportunities in the 21st Century. *Kashf Journal of Multidisciplinary Research* 1(05): 35-46.

Sutter, R. (2024) Professional Identify Formation: History, Practice and Challenges.

Swain, J. M., Spire, Z. D. (2020) The role of informal conversations in generating data, and the ethical and methodological issues. In *Forum Qualitative Sozialforschung/Forum: Qualitative Social Research* 21:1.

Taber, K. S., Billingsley, B., Riga, F., et al. (2015) English secondary students' thinking about the status of scientific theories: consistent, comprehensive, coherent and extensively evidenced explanations of aspects of the natural world–or just 'an idea someone has'. *The Curriculum Journal* 26(3): 370-403.

Tang, B. L. (2024) Publishing important work that lacks validity or reproducibility–pushing frontiers or corrupting science? *Accountability in Research*.

Taurino, A., Carlino, E. (2023) The Relevance of Building an Appropriate Environment around an Atomic Resolution Transmission Electron Microscope as Prerequisite for Reliable Quantitative Experiments: It Should Be Obvious, but It Is a Subtle Never-Ending Story! *Materials* 16(3): 1123.

Teymourifar, A. (2024) Balancing Innovation and Ethics: Integrating ChatGPT in Educational Environments for Enhanced Learning and Responsible Use.

Theron, F. (2024) An Overview of the Evolution of Diplomacy in Europe and Africa. *In* Key Issues in African Diplomacy: Developments and Achievements.

Thomke, S. H. (2020) *Experimentation works: The surprising power of business experiments*. Harvard Business Press.

Thomson, O. P., Martini, C. (2024) Pseudoscience: A skeleton in osteopathy's closet? *International Journal of Osteopathic Medicine* 52: 100716.

Thurik, A. R., Audretsch, D. B., Block, et al. (2024) The impact of entrepreneurship research on other academic fields. *Small Business Economics* 62(2): 727-751.

Tomić, V., Buljan, I., Marušić, A. (2024). Development of consensus on essential virtues for ethics and research integrity training using a modified Delphi approach. *Accountability in Research* 31(4): 327-350.

Traeger, A. C., Bero, L. A. (2024) Corporate Influences on Science and Health—the Case of Spinal Cord Stimulation. *JAMA Internal Medicine* 184(2): 129-130.

Udegbe, F. C., Ebulue, O. R., Ebulue, C. C., et al. (2024) AI's impact on personalized medicine: Tailoring treatments for improved health outcomes. *Engineering Science & Technology Journal* 5(4): 1386-1394.

Undheim, T. A. (2024) An interdisciplinary review of systemic risk factors leading up to existential risks. *Progress in Disaster Science*.

Uttley, L., Quintana, D. S., Montgomery, et al. (2023) The problems with systematic reviews: a living systematic review. *Journal of Clinical Epidemiology* 156: 30-41.

Van Thiel, G. S., Cole, B. J. (2024) Costs and Research Funding: Public and Private Sources. *In* Biologic Knee Reconstruction. CRC Press.

Varas, D., Santana, M., Nussbaum, M., et al. (2023). Teachers' strategies and challenges in teaching 21st century skills: Little common understanding. *Thinking Skills and Creativity* 48:101289.

Varpio, L., Paradis, E., Uijtdehaage, S, et al. (2020) The distinctions between theory, theoretical framework, and conceptual framework. *Academic Medicine* 95(7): 989-994.

Verburgt, L. M. (2024) Debating contemporary approaches to the history of science.

Volti, R., Croissant, J. (2024) *Society and technological change.* Waveland Press.

von Luepke, H., Neuhoff, K., Marchewitz, C. (2024). Bridges over troubled waters: Climate clubs, alliances, and partnerships as safeguards for effective international cooperation? *International Environmental Agreements: Politics, Law and Economics.*

von Schwarz, E. R. (2024a) *Ethics of Modern Stem Cell Research and Therapy.* Cham: Springer Nature Switzerland.

von Schwarz, E. R. (2024b). Ethical Issues. *In* Ethics of Modern Stem Cell Research and Therapy. Cham: Springer Nature Switzerland.

Vos, B. (2021) Integrated HPS? Formal versus historical approaches to philosophy of science. *Synthese* 199(5): 14509-14533.

Waltman, L., Kaltenbrunner, W., Pinfield, S., et al. (2023) How to improve scientific peer review: Four schools of thought. *Learned Publishing* 36(3): 334-347.

Wang, Y., Shao, H., Zhang, C., et al. (2023) Molecular dynamics for electrocatalysis: Mechanism explanation and performance prediction. *Energy Reviews* 2(3): 100028.

Watson, A., Chapman, R., Shafai, G., et al. (2023) FDA regulations and prescription digital therapeutics: Evolving with the technologies they regulate. *Frontiers in Digital Health* 5: 1086219.

West-Eberhard, M. J. (2024) Animal behaviour and the new natural history. *Animal Behaviour.*

Whewell, W. (2023) *History of scientific ideas.* BoD–Books on Demand.

Wie, S. H., Jung, J., Kim, W. J. (2023) Effective vaccination and education strategies for emerging infectious diseases such as COVID-19. *Journal of Korean Medical Science* 38(44).

Wiemken, T. L., Kelley, R. R. (2020) Machine learning in epidemiology and health outcomes research. *Annu Rev Public Health* 41(1): 21-36.

Willis, L. D. (2024) The Peer Review Process. *Respiratory Care* 69(4): 492-499.

Williams, P. R., Paustenbach, D. J. (2024). Risk characterization. *Human and Ecological Risk Assessment: Theory and Practice* 1: 263-331.

Wu, L., Yi, F., Bu, Y., et al. (2024) Toward scientific collaboration: A cost-benefit perspective. *Research Policy* 53(2): 104943.

Xiao, X. (2024) Let's verify and rectify! Examining the nuanced influence of risk appraisal and norms in combatting misinformation. *New Media & Society* 26(7): 3786-3809.

Yeo-Teh, N. S. L., Tang, B. L. (2024) Research data mismanagement–from questionable research practice to research misconduct. *Accountability in Research* 31(6): 706-713.

Youvan, D. C. (2024) Jane Goodall's Legacy: Bridging Science, Conservation, and Humanity.

Zemla, J. C., Sloman, S. A., Bechlivanidis, C., et al. (2023). Not so simple! Causal mechanisms increase preference for complex explanations. *Cognition* 239: 105551.

Zhong, S., Zhang, K., Bagheri, et al. (2021) Machine learning: new ideas and tools in environmental science and engineering. *Environmental Science & Technology* 55(19): 12741-12754.

Ziman, J., Ziman, J. M. (1991) *Reliable knowledge: An exploration of the grounds for belief in science.* Cambridge University Press.

Zuccarini, G., Malgieri, M. (2024). Modeling and representing conceptual change in the learning of successive theories: The case of the classical-quantum transition. *Science & Education* 33(3): 717-761.

www.ingramcontent.com/pod-product-compliance
Lightning Source LLC
Chambersburg PA
CBHW071734120626
46550CB00002B/514